199 Color TV Troubles & Solutions

by R. L. Goodman

TAB BOOKS
Blue Ridge Summit, Pa. 17214

FIRST EDITION

FIRST PRINTING—APRIL 1972

Copyright © 1972 by TAB BOOKS

Printed in the United States
of America

Reproduction or publication of the content in any manher, without express permission of the publisher, is prohibited. No liability is assumed with respect to the use of the information herein.

International Standard Book No. 0-8306-2595-X

Library of Congress Card Number: 71-188708

Preface

This handbook is for the technician who makes his living in the TV service business. The best troubleshooting system for any color TV chassis is to use a logical procedure that will lead you right to the malfunction in the shortest possible time. However, certain types of recurring troubles have shown up in the past and knowing about them can cut many frustrating hours from your limited troubleshooting time. This is the reason for compiling this "case-history" book in order for you to have a handy reference of these past troubles and solutions. Just don't get into the "rut" of always relying on the "case-book-tips," but use your own good logical troubleshooting procedures. After all, you have to repair the chassis with your own mind and hands.

These case-histories were gleaned from the author's many years of personal experience at the color TV service bench, and some are from TV manufacture service tips and circuit modifications. The symptom and solution is presented with only the "meat" necessary for you to solve the problem speedily.

Some of the following "case-history" problems not only show the solutions but a troubleshooting guide is given to help solve future "tough-dog" TV chassis troubles.

The troubles are listed by set symptoms in the contents and according to chassis numbers in the index. Many TV circuits, regardless of who makes it, do about the same job. If you have a Zenith color set problem, it's possible you may find the solution under a Motorola or Admiral case history. In many cases they will at least get you started in the right direction.

I am sure the case-histories in this book will be very useful to you.

Robert L. Goodman

Contents

1	ADMIRAL COLOR MONITOR—No color	17
2	ADMIRAL G11—No color	18
3	ADMIRAL G12—Poor purity	19
4	ADMIRAL G13—Unstable horiz. hold	20
5	ADMIRAL G13—High voltage will not adjust	21
6	ADMIRAL G13—Weak or no color	22
7	ADMIRAL G13—No picture	23
8	ADMIRAL H10, H12 & K15—Dark vertical bars approx. one inch wide on left side of screen	24
9	ADMIRAL H10, H12—Poor focus, circuit modification	24
10	ADMIRAL H12—Narrow, out-of-focus picture	26
11	ADMIRAL H12—Loss of blue or red	27
12	ADMIRAL H12—Color hue shift	28
13	ADMIRAL H12—Convergence problems	29
14	ADMIRAL H12—Horizontal squeal, no picture, no high voltage	30
15	ADMIRAL H12—AGC overload	31
16	ADMIRAL H12—No sound	32

17	ADMIRAL 4K10—No high voltage	33
18	ADMIRAL 4K10—AGC circuit problems	34
19	ADMIRAL K10—Audio system troubles	35
20	ADMIRAL K10—Distorted sound	36
21	ADMIRAL K15—Erratic color lock	37
22	ADMIRAL K15—No HV control	38
23	ADMIRAL K15—Raster impurity	39
24	G.E.—Raster weaving horizontally	40
25	G.E. C1—Damper "Hausen" lines	41
26	G.E. C1—Focus tracking network arc and raster bloom	42
27	G.E. C2—Vertical sweep troubles	44
28	G.E. C2—Arcing, streaks across picture	45
29	G.E. CB—No color	46
30	G.E. CB—Dull, washed-out picture	47
31	G.E. H3—No raster, no high voltage	48
32	G.E. H3—No color or weak color	49
33	G.E. H3—Audio distortion	50
34	G.E. H3—Piecrust (slight) horizontal AFC problem	51
35	G.E. H3—No horizontal sweep or high voltage	51
36	G.E. H3—Vertical jitter	53

37	G.E. H3—No vertical sweep	53
38	G.E. KC—Intermittent focus symptom	54
39	G.E. KD—Picture pulls in from sides of screen	55
40	G.E. KE—Vertical color stripe on screen	56
41	G.E. KE—Vertical size & retrace lines	57
42	G.E. KE—Weak video	58
43	G.E. KE—Tweet on Channel 5 or 11	59
44	G.E. KE—Raster shading problems	60
45	G.E. KE—Horizontal pull	61
46	G.E. KE—Picture bending or snow	62
47	G.E. KE—Video troubles	63
48	G.E. KE—Raster pulled in from sides	65
49	G.E. KE—Lines across screen	65
50	MAGNAVOX—Failure of color detector diodes	66
51	MAGNAVOX T924—Picture top hook	67
52	MAGNAVOX T933—Critical horizontal locking	68
53	MAGNAVOX T933—Critical color sync	69
54	MAGNAVOX T933—Overall tinted raster, low color saturation	69
55	MAGNAVOX T935—Vertical jitter	71
56	MAGNAVOX T938—Weak color	72
57	MAGNAVOX T908-T915—Unstable picture	73

58	MAGNAVOX T919—Low brightness	74
59	MAGNAVOX T936—Weak video	75
60	MAGNAVOX T931, T933 & T940—No color or poor color sync	76
61	MAGNAVOX—Remote control receiver model 704058—station skipping. May be intermittent	77
62	MOTOROLA TS-908—Vertical white drive line	79
63	MOTOROLA TS-908—Picture side pull	80
64	MOTOROLA TS-908—Low picture brightness	81
65	MOTOROLA TS-914—Circuit modification for insufficient brightness range	82
66	MOTOROLA TS-914—No video information	83
67	MOTOROLA TS-914-918—Picture not clear, poor peripheral convergence	84
68	MOTOROLA TS-914-918—Screen will not light, no HV at anode cup	84
69	MOTOROLA TS-918—Color killer critical to adjust	86
70	MOTOROLA TS-918—No color	86
71	MOTOROLA TS-918—No sound	88
72	MOTOROLA TS-918—Focus problem	89
73	MOTOROLA TS-918—Gray vertical bars	90
74	MOTOROLA TS-918—Low brightness level	91

75	MOTOROLA TS-918—Blurred video or low brightness	92
76	MOTOROLA TS-915-919—Picture blue and not clear	93
77	MOTOROLA TS-915-919—Narrow picture	94
78	MOTOROLA TS-915-919—No high voltage	95
79	MOTOROLA TS-924—Poor resolution	96
80	MOTOROLA TS-934A—Hum bar	97
81	PHILCO 19 FT 20 & OTHERS—Vertical ringing (bars on the left side of screen)	99
82	PHILCO 19FT20—Vertical buzz in the sound	100
83	PHILCO 19FT20—Picture lacks detail	101
84	PHILCO 17KT50—One color missing	102
85	PHILCO 17KT50—Vertical jitter	103
86	PHILCO 20KT40—Color lacking red	104
87	PHILCO 20KT40B—Critical horizontal sync	106
88	PHILCO 20KT41B—Loss of color	107
89	PHILCO 12L80 & 13L80—Poor sync	108
90	PHILCO 15M90—Narrow picture	109
91	PHILCO 15M90—Picture pulled in from top & bottom	110
92	PHILCO 15M91—Color bars through picture	110

93	PHILCO 17MT80—Vertical jitter or poor interlace	112
94	PHILCO 18MT10-B—Hash or weak negative picture	113
95	PHILCO 16QT85—Picture hash	114
96	PHILCO 16QT85-A—Dark bars or bands on raster	115
97	PHILCO 17QT82—No raster or high voltage	116
98	PHILCO 18QT85-A—No color or random color	117
99	PHILCO 19QT85-B—Loss of color or improper sync	118
100	PHILCO 19QT87-B—"Odd Ball" mixed up color screen	119
101	PHILCO 19QT87-B—A shifty picture	120
102	PHILCO 20AT88-B—Poor horizontal sync or horizontal hold	121
103	PHILCO 20QT90-B—No color	122
104	RCA CTC15—Intermittent raster fade-out	124
105	RCA CTC16X—Distorted or compressed scanning lines	125
106	RCA CTC16X—Blurred picture & low high voltage	126
107	RCA CTC17X—Color confetti on B&W picture	127

108	RCA CTC21—Dark screen, no raster	128
109	RCA CTC21—Blurry picture	129
110	RCA CTC21 & OTHERS—Focus problems	130
111	RCA CTC21—Hum or buzz in audio	132
112	RCA CTC22—Vertical & horizontal sync instability	133
113	RCA CTC22—Scalloped picture	134
114	RCA CTC22—No vertical sweep	135
115	RCA CTC24—No color	136
116	RCA CTC24—Raster will not focus	137
117	RCA CTC25A—Sound troubles	138
118	RCA CTC25X—No video or weak & washed-out picture	139
119	RCA CTC25X—No raster	140
120	RCA CTC25X—No HV or raster	141
121	RCA CTC25X—Poor picture lock-in	142
122	RCA CTC27X—Gray scale shift	143
123	RCA CTC31—Color fades out	144
124	RCA CTC31—No color	145
125	RCA CTC31—Washed out picture	146
126	RCA CTC31A—Low brightness	147
127	RCA CTC38—Bright, washed-out picture	148

128	RCA CTC38—Weak color & loss of color sync	149
129	RCA CTC40—Excessive brightness	150
130	RCA CTC40—Video "smear" and loss of sync	151
131	RCA CTC40, 47—Piecrust or geartooth effect	152
132	RCA CTC44-47—Hum bars and-or degaussing interference	154
133	RCA CTC46-49—High voltage troubles	155
134	RCA CTC52—Picture "hum-bar" interference	156
135	ZENITH 14A9C29Z—Color fades out	157
136	ZENITH 14A9C50—Diagonal lines from corner of screen	158
137	ZENITH 14A9C50—Streaks across screen	159
138	ZENITH 14A9C50—No picture or sound	160
139	ZENITH 14A9C50 USING DEMODULATOR IC "CHIP"—Poor color reception	161
140 141 142	ZENITH 14A9C50—Triple trouble symptoms may be found in this color TV	162
143	ZENITH 14A9C51—Loss of blue	164
144	ZENITH 14A9C51—Shimmering picture	165
145	ZENITH 14A9C51—Vertical roll	166
146	ZENITH 14A9C51—No color	167
147	ZENITH 14A10C19—Brightness level varies	168

148 ZENITH 14A10C27Z—High line voltage areas   169

149 ZENITH 12A12C52—No sync & loss of color   170

150 ZENITH 12A12C52—No raster   172

151 ZENITH 12A12C52—Loss of high-voltage   172

152 ZENITH 12A12C52   173

153 ZENITH 12A12C52   174

154 ZENITH 12A12C52, 20X1C36, & OTHERS—
     Color picture moire   175

155 ZENITH 4B25C19—Picture "ghost"   177

156 ZENITH 4B25C19—Loss of color sync   178

157 ZENITH 4B25C19—No raster and-or
     loss of brightness   178

158 ZENITH 4B25C19—Screen dim or black-out   180

159 ZENITH 4B25C19—Intermittent sound condition   181

160 ZENITH 4B25C19—Blue missing in picture   182

161 ZENITH 4B25C19—Screen too bright   183

162 ZENITH 12B14C50—No raster   184

163 ZENITH 12B14C52—Very weak color   185

164 ZENITH 12B14C52—Low level color   186

165 ZENITH 12B14C52—Low volume and distorted
     audio   187

166 ZENITH 12B14C52—A blue blood color receiver   188

167	ZENITH 12B14C52—Washed out color	189
168	ZENITH 12B14C52—Color trouble	190
169	ZENITH 14B9C50—Bars across screen	191
170	ZENITH 14CC14—No color	192
171	ZENITH 25NC37—Silicon rectifier failures	193
172	ZENITH 20X1C36 COLOR—A short circuit problem	194
173	ZENITH 20X1C38—Color sync trouble	195
174	ZENITH 15Y6C15—Convergence trouble	196
175	ZENITH 20Y1C38—Hue control varies color	198
176	ZENITH 20Y1C48—Picture convergence drift	199
177	ZENITH 20Y1C50—Weak color	200
178	ZENITH 14Z8C50—HV regulation trouble	201
179	ZENITH 14Z8C50—Left side of raster green	202
180	ZENITH 14Z8C50—Dark horizontal hum bar	203
181	ZENITH 14Z8C50—Vertical roll	204
182	ZENITH 14Z8C50—Tracking modification	205
183	ZENITH 16Z7C19—Wrong colors	206
184	ZENITH 16Z7C19—Color comes and goes	206
185	ZENITH 16Z7C19—Colors not correct	207
186	ZENITH 16Z7C50Z—Ringing	208
187	ZENITH 16Z7C50—VHF tuner drift problem	210

188	ZENITH 16Z7C50Z—Color "drop out"	212
189	ZENITH 16Z7C50Z—Not enough vertical sweep	213
190	ZENITH 16Z7C60Z—Erratic AGC	214
191 192	ZENITH 16Z7C50—Picture smear and raster shrinkage	216
193	ZENITH 16Z8C19—Picture interference patterns	217
194	ZENITH 16Z8C19—Hum bars through picture	219
195	ZENITH 16Z8C50—Picture bends and pulls	220
196	ZENITH 16Z8C50—Drive lines and AGC trouble	221
197	ZENITH 20Z1C37—Vertical shading	
198	ALL BRANDS OF COLOR RECEIVERS—Intermittent circuit connections & circuit board leakage	223
199	LOTS OF SNOW ON THE SCREEN	223

ADMIRAL COLOR MONITOR
No color

For the Admiral color chassis using the "color monitor" circuit (Fig. 1), which is a phase-control type circuit to improve flesh tones, complaints of no color may be due to a color-monitor circuit fault. The chroma signals go through the color-monitor chassis regardless of which position the switch is set in. Input is to pin 1 of the three-pin plug, through a .01-mfd capacitor to the base of the chroma driver transistor. The signal then goes to the color amplifier transistor, and thus to pin 2 of the plug.

Use your scope to check for the presence or absence of this color signal. Feed a color-bar signal into the antenna of the set and check for the amplitude of the "rocker" pattern on pins 1 and 2 of the plug socket combination. Use a demodulator probe, because this is an RF signal. Now, pull the plug from the color monitor chassis, and short pins 1 and 2 of the plug together. This will let the color information go onto the screen. If the chroma signal does not go through the color monitor circuitry, check the DC voltages, transistors, and all the diodes. Refer to the schematic for the correct DC voltages.

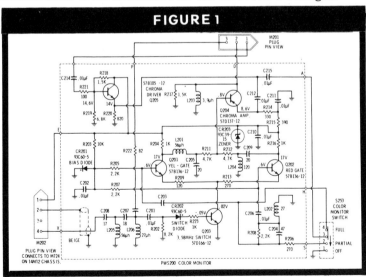

FIGURE 1

2 ADMIRAL G11 CHASSIS
No color

We have repaired several of these G11 chassis that had this same symptom and circuit fault. The symptom may be no color or weak and intermittent color reception. Refer to the color killer schematic in Fig. 2 as to the possible component fault that will cause this lack of color. The defective part is C522, a .1 mfd capacitor in the grid circuit of V504A color killer stage. The failure of this capacitor causes a greater negative voltage to be produced at the grid (pin 9) of the color killer stage which will bias off the 2nd bandpass amplifier. Replace with a .1 mfd, 600 W.V.D.C. rated capacitor. Check and adjust color killer control R702 if needed.

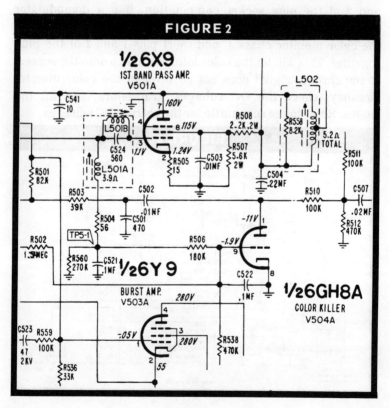

ADMIRAL G12 CHASSIS
Poor purity

3

If some difficulty is experienced in obtaining good picture purity, check the wiring of the (L109) filter choke. Refer to the power supply schematic in Fig. 3. Some of these chokes have two yellow leads and they may have been installed incorrectly. Reversing the lead connections may correct the purity problem by properly polarizing the filter choke windings. In later chassis production, a filter choke was installed that used a black lead for the 400-volt supply and a yellow lead for the 415 B+ voltage supply source.

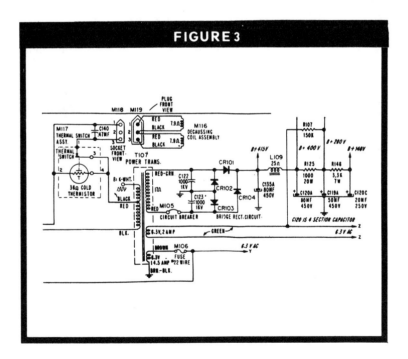

FIGURE 3

ADMIRAL G13 CHASSIS
Unstable Horizontal Hold

This color TV receiver had developed an unstable horizontal hold condition. For this symptom make the following quick checks. With a jumper lead, short test point TP4-5 to ground, thus, removing the correction voltage from the AFC. With the horizontal hold control we should be able to lock in a picture, but it will have a slight drift. With the jumper lead removed the picture should now sync in and lock. If you cannot lock in a picture with the hold control, the problem is in the oscillator. If the oscillator is normal, the fault is in the control circuit. The following components are the most likely ones to fail in the oscillator; C442, C443, C444, C445 (not shown in Fig. 4), or C711B, and the horizontal hold coil. In the horizontal control circuit (Fig. 4) check C418, C436, C437, C439, and dual-diode CR401.

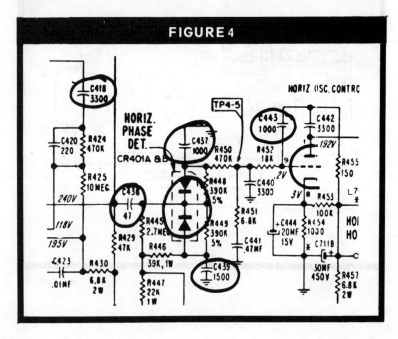

FIGURE 4

ADMIRAL G13 CHASSIS
High voltage would not adjust

5

This color chassis HV control had no effect and the HV measured at more than 30 KV. It was also noted that the cathode current for the 6KD6 was excessive. A defective component was suspected in the regulator diode circuit shown in Fig. 5. Check for a negative voltage at the diode. It should be approximately -120v. The regulator diode can short or open, and may also become intermittent. Also, be on the lookout for poor solder connections to the PC board. Either case will cause excessive HV and current. Check for a shorted C449 or an open C451. The HV regulator diode must receive a pulse from a winding on the horizontal sweep transformer. Check the continuity of this winding. If found defective, replace regulator diode (CR402) with Part No. 93B60-3 and capacitor C449 (.01 mfd).

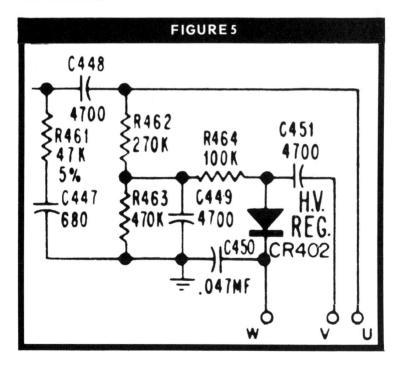

FIGURE 5

ADMIRAL G13 CHASSIS
Weak or no color

Replace defective capacitor C521. This .1 mfd capacitor may check out good with an ohmmeter. The voltage from the killer stage is biasing off the control grid of the second bandpass amplifier stage. If the screen voltage on the first color bandpass amplifier is low and does not change when the tube is removed, check for a faulty C503 capacitor. This may be a green disc capacitor and is very likely to fail when used for this circuit application. Replace the .01 mfd disc capacitor. Note schematic in Fig. 6.

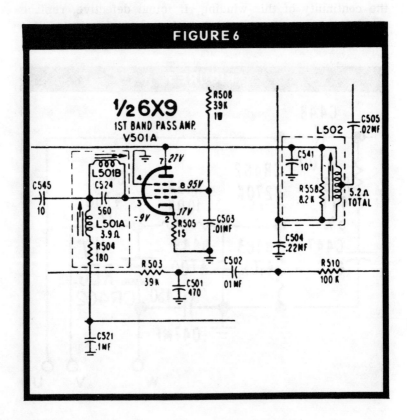

ADMIRAL G13 CHASSIS
No picture

7

This chassis came into the shop with no picture (black screen) and the HV measured zero. The trouble appeared to be in the horizontal sweep output section. Many voltage checks and tests were made before we found this defect. Finally, a defective horizontal centering control was discovered. This is a 10-ohm control shown as R734 in the horizontal sweep circuit of Fig. 7. This control may also be burnt, but you can smell this trouble. In some chassis this control will be defective, but you will have high voltage with only a thin, bright vertical line in the center of the screen. Replace the horizontal centering control (R734) and check for correct HV at the CRT anode.

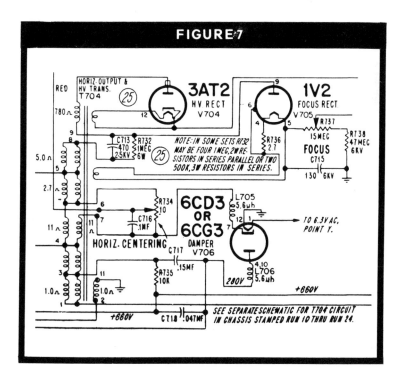

8 ADMIRAL H10, H12 AND K15 CHASSIS
Dark vertical bars approximately one inch wide on left side of screen

Caused by an improper lead dress condition. Check the lead dress of the CRT wiring. The grid and cathode leads should be in a fishpaper retainer, dressed away from the horizontal sweep transformer. Reroute the "blue" lead from the 1st video amplifier to the chroma section. The lead connects point "MM" on PWS400 to point "CC" on PWS500. This lead will be found with many other leads in a bundle. Unsolder the lead at the chroma board, move it away from the other leads, and reroute the lead through the center area of board PWS400.

9 ADMIRAL H10, H12 SERIES CHASSIS
Poor focus, circuit modification

In some of these chassis, the focus control is wired as a rheostat instead of a potentiometer. A 1.8 meg resistor (R747) has also been added (see Fig. 8). This modification should be made whenever you replace the focus control in an H10 or H12 chassis. If the focus range is not adequate after this modification, try a lower value for the added resistor (R747), 1.0 meg, for instance, but keep the value as high as possible.

Notice that control part numbers 75C108-1, -2 and -3, are electrically interchangeable, differing only mechanically. All three can be used for replacement. The -1 does not have a shaft. While the -2 and -3 both have shafts but slightly different terminal lug arrangements.

FIGURE 8

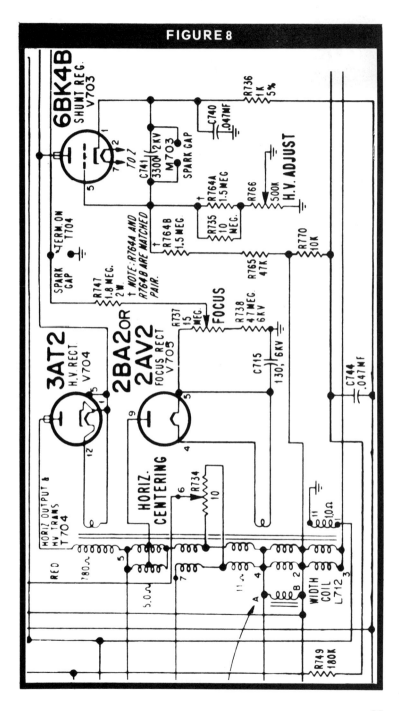

10 ADMIRAL H12 CHASSIS
Narrow, out-of-focus picture

This color chassis displayed a picture with insufficient width. The raster was out of focus and the 6KD6 H.O.T. was pulling excessive current. Of course, you must change all of the horizontal sweep section tubes. Next, correct by adjusting the horizontal efficiency coil, as this coil may be turned out too far, giving the above results. Connect the current meter and adjust coil L713 until a dip of the least amount of current is noted on the meter. This adjustment should be made whenever a routine current check is performed. Examine horizontal efficiency coil L713 for any burned marks. If in doubt, change it. Refer to Fig. 9.

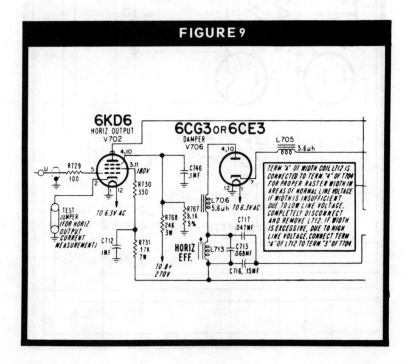

FIGURE 9

ADMIRAL H12 CHASSIS
Loss of blue or red

Make the following checks: Test for failure of phase coils L506 and L507 (replacement Part No. 73B55-26). A breakdown of the spark gaps in the CRT leads is a possible cause for failure of these coils.

Two types of spark gaps have been used. Late production sets use a "Gap-Cap," while early production sets use a twin-lead type gap. Some of the early spark gaps have been found with the insulation protruding past the wire ends. The ends of the wires must be exposed in order for the gap to operate. Trim the twin-lead insulation until the ends of the wire are exposed slightly beyond the insulation. The exposed ends must not be moved from their original spacing. Replacement of either L506 or L507 without checking the spark gaps may result in repeated coil failures. The schematic for this color section circuit trouble is shown in Fig. 10.

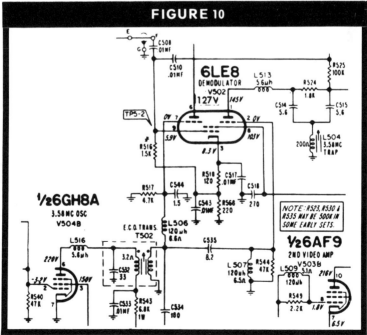

12

ADMIRAL H12 CHASSIS
Color hue shift

This color set was working fine, except the flesh tones would shift drastically with a camera change. Colors would go from normal to purple from camera to camera (some color programming may seem to be normal), in which case the voltage on the grid (pin 1) of the burst amplifier (6AF4) will have considerably more positive voltage than shown on the schematic. The screen may also go green for an instant. This shifty color problem was caused by a leaky capacitor C701 with a value of 120 pf, shown in the schematic of Fig. 11. If the color fades out on some channels, but other TV stations are good, try readjusting the color killer control for all stations received.

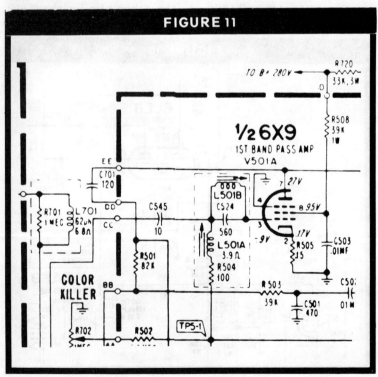

ADMIRAL H12 CHASSIS
Convergence problems

13

On some chassis, T601 and R601, have overheated. Both components are located on the convergence panel (Fig. 12). In later production, R601 has been changed from (120 ohms, 2 watts) to (120 ohms, 3 watts), and a 30-ohm resistor, R619, added in series with R601. For field service replacement, use the 120-ohm, 3-watt unit and add the 30-ohm, 5-watt resistor in series with either side of R601 by cutting the foil and mounting the resistor on the foil side of the PC board.

If R601 or T601 overheats, check the adjustment of the blue shaper coil. Remember that the blue shaper coil is not a convergence adjustment and must be set according to the manufacturers instructions. Should it be necessary to change T601 or the other coils on this board, be sure to use correct replacement coils. The 93D175 and 94D305 part numbers have been used. Because the basic configuration of the coils is different, they cannot be interchanged.

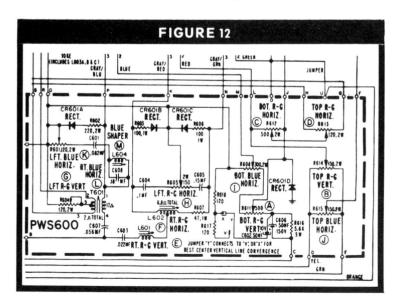

FIGURE 12

ADMIRAL H12 CHASSIS
Horizontal squeal, no picture, no high voltage

The most obvious symptom for this set was a loud horizontal squeal. There was no picture or high voltage. A few voltage checks in the horizontal oscillator and sweep output stage were about normal; however, the drive voltage was low. With a few scope checks, we learned that the oscillator frequency was too high. The trouble was caused by a faulty C711 capacitor shown in the horizontal oscillator circuit, Fig. 13. This is a 30 mfd, 450v filter capacitor.

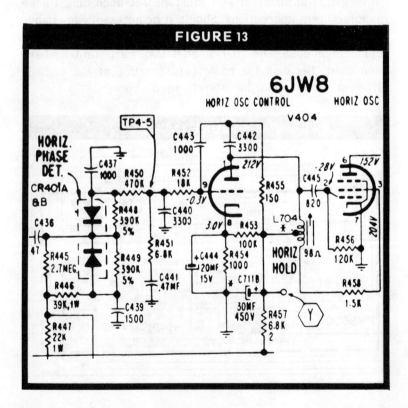

ADMIRAL H12 CHASSIS
AGC overload

15

The picture and sound were good on the weak signal TV channels, but the picture indicated circuit overload on a strong TV station. All of the voltages shown in the sync and AGC circuit of Fig. 14 were checked and found correct. The trouble was located as a leaky C416 capacitor. This .001 mfd capacitor feeds the horizontal keying pulse to the plate of the AGC stage. After the capacitor was replaced and the AGC control adjusted, all stations were received clearly.

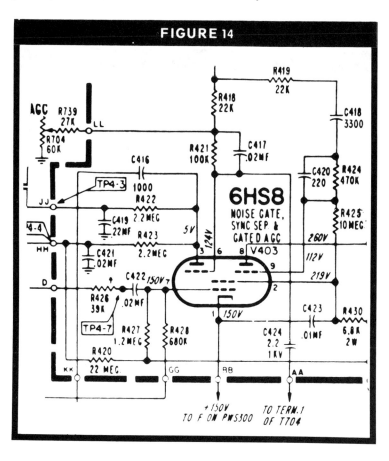

FIGURE 14

16 — ADMIRAL H12 CHASSIS
No sound

This set had a good color picture, but no sound came from the speaker. A few voltages were checked in the sound section and the trouble was narrowed down to the detector stage. A few more checks revealed that quadrature sound detector coil L404 was open. This may only be a cold solder joint in the coil, in which case it can be repaired. If not, replace L404, shown in the sound circuit of Fig. 15. Then align the sound stage.

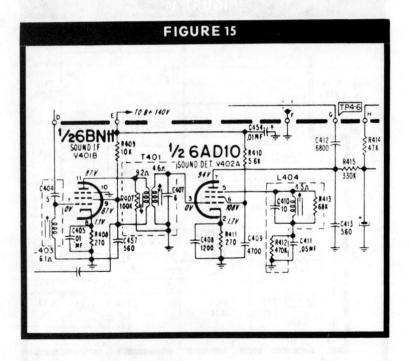

FIGURE 15

ADMIRAL 4K10 CHASSIS
No high voltage

17

Remove the 25CK3 damper tube and measure for a short between the plate and cathode pins of this socket. If there is a short, check for a faulty CH65 (82pf, 5KV) capacitor. Note horizontal sweep circuit in Fig. 16.

Also check the current of the horizontal output tube, and don't run the CRT screens too high. With the screen, contrast and brightness controls turned up, this chassis will draw over 230 ma of current and "wipe-out" the sweep transformer. For a current check, use a burned-out bell fuse. Remove the plastic cap and clip two leads from the bell fuse to the current meter. To remove the HV rectifier tube (3DF3), you must first remove the horizontal output and damper tubes from their sockets before you can open the HV door cover. Should the 4K10 chassis keep "blowing" the bell fuse, suspect an intermittent short in the horizontal output or damper tube. Monitor the current meter and reset the screen controls if necessary.

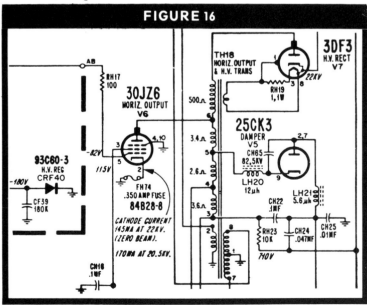

18

ADMIRAL 4K10 CHASSIS
AGC circuit problems

If the picture appears to have an AGC overload symptom, check the voltage on the base of Q8 (AGC amp). If it is negative, try applying an external bias voltage. If normal reception is restored, check for an open electrolytic capacitor, CB50 (100 mfd, 10v). Refer to the AGC section schematic in Fig. 17. Should CB50 short, adding bias will have no effect. The set will produce a washed-out picture or a blank raster.

When the picture is intermittently weak or snowy, check the setting of the AGC delay control (RH36). The control (RH36) may also be defective (erratic). Also suspect, then check, transistors Q7, Q8 and Q9.

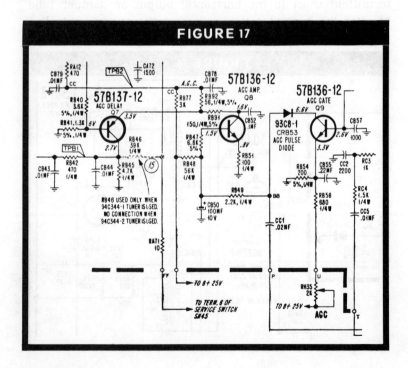

FIGURE 17

ADMIRAL K10 CHASSIS
Audio system troubles

19

For a no sound or distorted sound condition, check Q3, Q4, and Q5 in the IF sound section. See sound section circuits in Fig. 18. Check the condition of diodes CRB25 and CRB26. Check CRB88 for correct voltage regulation to Q5. Also check CB12 (4700 pf) for very sharp sound tuning.

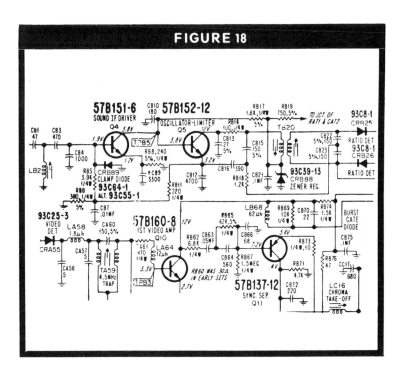

FIGURE 18

20 ADMIRAL K10 CHASSIS
Distorted sound

If you have a set with distorted sound, check the voltages on Q4 (sound IF driver), Q5 (oscillator limiter), Q6 (audio preamp) and Q22 (audio output) as shown in the sound circuits of Figs. 18 and 19. If the voltages are OK, check the sound alignment. May also be a defective audio transistor (Q4). Check for faulty capacitors CB13 (27pf), CB22 and CB23 (150 pf), or the sound detector (CRA43, not shown) for leakage.

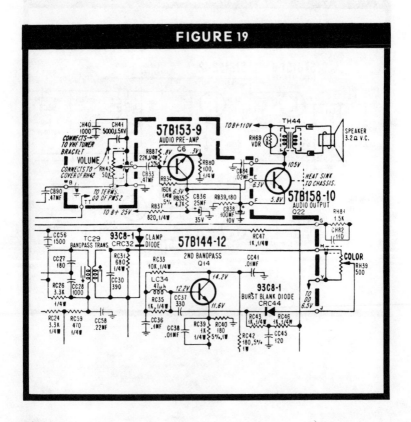

FIGURE 19

ADMIRAL K15 CHASSIS
Erratic color lock

21

The color sync on this chassis was erratic, and when switching from channel to channel the problem appeared intermittently. The set had a clear black and white picture with good audio. We suspected a cold solder joint in the burst amplifier or color oscillator circuit board section. When the color burst transformer was moved we noted that the trouble would appear. After some more detective work, the problem was located inside the T501 burst transformer can. Capacitors C527 and C528 in the burst transformer were defective. Notice the color burst system schematic shown in Fig. 20. You may want to replace the complete transformer unit.

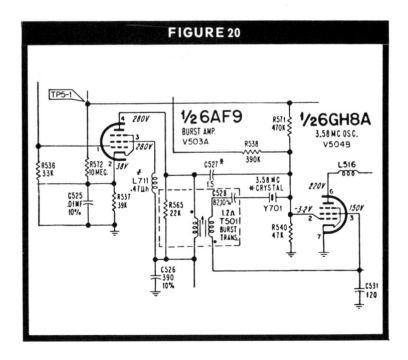

FIGURE 20

22 ADMIRAL K15 CHASSIS
No HV control

Everything seemed to be working normally in this chassis, but the high voltage was excessive. There was very little HV regulation and the HV adjust control had no effect. The CRT anode voltage measured over 30KV and the cathode current of the 6KD6 was excessive. We now looked for a defective component in the regulator diode circuit shown in the schematic of Fig. 21. There should be a negative voltage at diode CR402, approximately -120 volts. The regulator diode may short out or open up. Either of these conditions will cause excessive HV and H.O.T. cathode current. Also, check for a shorted C449 (.01 mfd) or open C451. The diode must receive a horizontal gating pulse from a winding of the sweep transformer. Use your scope to check for the presence of this gating pulse. If the diode is defective, replace it with Admiral part number 93B60-3.

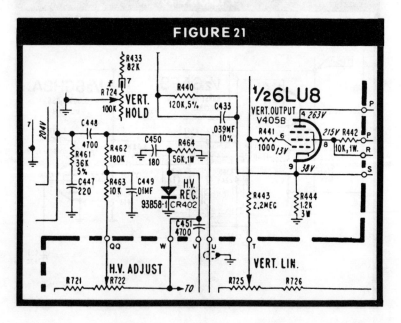

FIGURE 21

ADMIRAL K15 CHASSIS
Raster impurity

The left side of the raster on this set had become almost a solid, single color and looked like a purity problem. A faint, wide vertical bar may also be noticeable in the raster. Moving the grid leads to the picture tube would change the intensity on the left side of the raster.

Use the schematic in Fig. 22 and the following tips to locate this trouble. Look for an open C513, C516 or C520 (.047 mfd) capacitors. The color of the vertical bar or shading will indicate which capacitor. For example, C520 would produce green, etc. Also, checking the individual fields will reveal which color is affected.

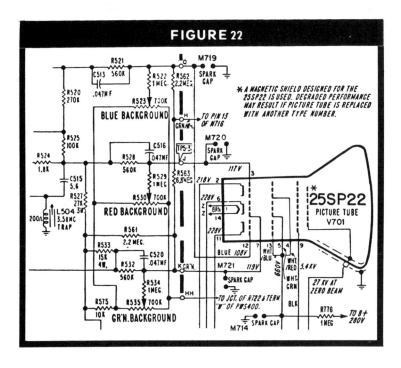

FIGURE 22

24 GENERAL ELECTRIC CHASSIS
Raster weaving horizontally

Raster weaves (pulls) horizontally at a 60-Hz rate. Probably caused by leakage in the damper tube (19CG3) between the filaments and pin 3. Pin 3 of the tube is not used, but pin 3 of the socket is used as a tie point for C276, C274, and R251. The resistor (R251) furnishes the reference signal for the horizontal AFC circuit. The 60-Hz signal thus modulates the reference signal and causes the weave. Replace the 19CG3 tube. The later production chassis use a terminal strip for this tie point in place of pin 3.

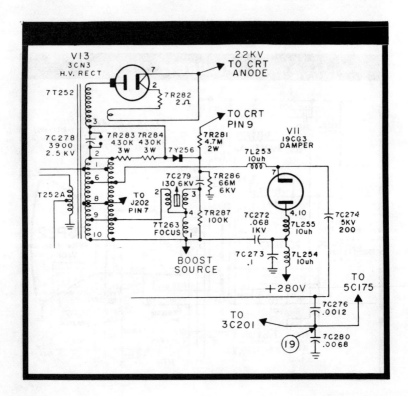

GENERAL ELECTRIC C1 CHASSIS
"Damper Hausen" lines

25

Some of these GE chassis will exhibit a watery line at the extreme left of the screen under weak signal conditions. This is caused by an oscillation in the damper circuit. The oscillations or "damper hausen" can be eliminated by connecting an RF choke in series with the damper tube plate, pin 4 of the 19CG3 tube. Later production chassis are equipped with this choke. Refer to the drawing in Fig. 23 for the modification to be made.

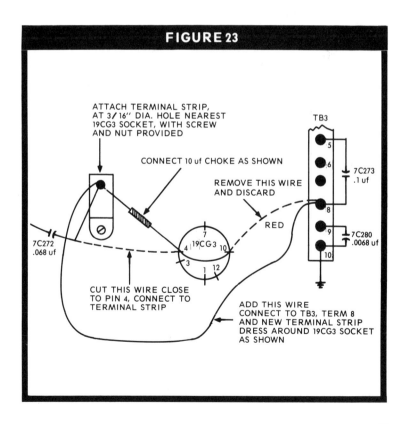

26
GENERAL ELECTRIC C1 CHASSIS
Focus tracking network arc and raster bloom

Blooming raster problems are normally associated with high voltage defects, but they can also be caused by excessive loading on the high voltage system. Refer to Fig. 24.

In C1 chassis receivers, excessive CRT beam currents can create a large voltage drop across the focus tracking network resistors, 7R283, 7R284, causing the spark gap to arc continuously. This lowers the CRT second anode voltage so that the raster blooms.

These excessive beam currents are the result of improper CRT element voltages—the bias voltage between the control grids and the cathodes being the most critical. The control grids are normally 100 to 150 volts negative with respect to the cathodes. (The actual voltages are determined by the brightness control.) If this grid-to-cathode bias should drop to 50 volts or less, heavy beam currents will flow.

A reduction in the grid-to-cathode bias is most likely caused by a decrease in cathode voltage relative to chassis ground. Since the video amplifier is DC coupled from the video detector to the CRT cathodes, any of the following defects could cause the CRT cathode voltages to decrease:

1. Plate to cathode short in V5A
2. V5A cathode to chassis short
3. Shorted 4C179
4. Shorted Q304
5. Shorted Q301
6. Open 3R169

Should you service a C1 chassis with a focus tracking arc and raster bloom, be sure to check the CRT bias voltages. This will help you to determine whether the defect is in the high voltage circuit or in the video amplifier circuit.

FIGURE 24

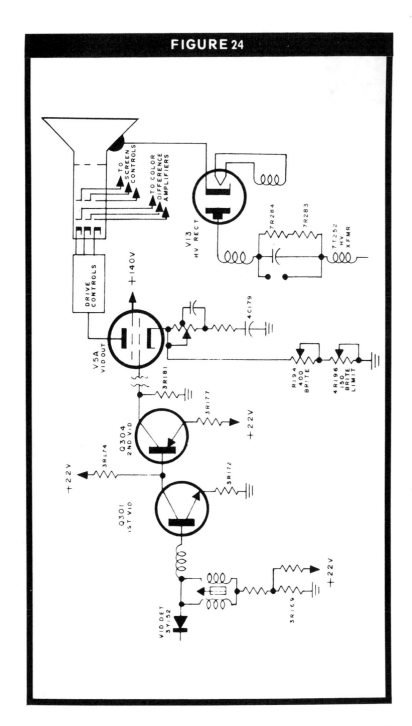

27 GENERAL ELECTRIC C2 CHASSIS
Vertical sweep troubles

With no vertical sweep, check 2R238, a 330-ohm 2-watt resistor; it has probably opened. Be sure and check the load side of the vertical sweep transformer for any possible short circuit. (Fig. 25)

If the vertical sweep is insufficient and cannot be adjusted with the height and linearity controls, the probable fault is an open 5C216, a 50-mfd 150v capacitor. Check or replace.

In cases of picture foldover at bottom of screen, suspect transformer 7T201. It may have decreased in inductance or have some shorted turns. Neither trouble would be found with an ohmmeter. The only sure bet is to substitute a new 7T201 vertical sweep output transformer.

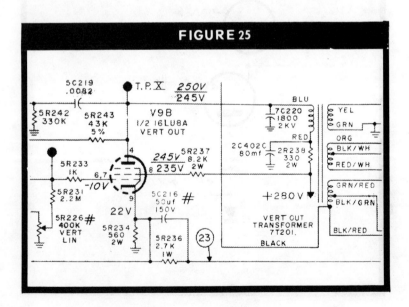

GENERAL ELECTRIC C2 CHASSIS
Arcing, streaks across picture

28

Arcing in the focus tracking network (Case History 26) may be due to a shorted 8AL9 (see Fig. 26) or capacitor 4C179. The contrast or brightness control wiring or a terminal may have become shorted to ground. Also, video transistors Q301 or Q304 may short and cause this same arcing. As with Case History 26, the result of any of the above troubles is excessive CRT beam current, which causes a heavy voltage drop across the focus tracking resistors.

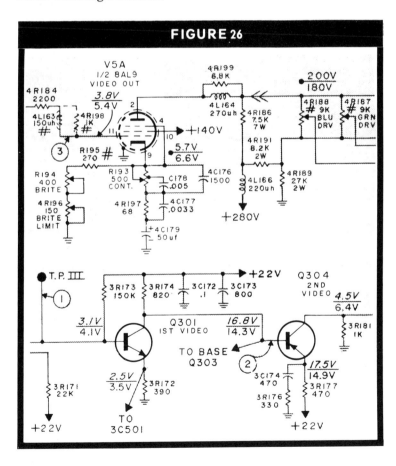

29 GENERAL ELECTRIC CB CHASSIS
No color

This GE color set had a good black-and-white picture, but there was no color at all. In this chassis the color burst signal rings the 3.58-MHz crystal and there is no color killer circuit or control. The color circuits for this set are shown in Fig. 27.

The oscilloscope was used to probe for incorrect waveshapes in the color stages and it was noticed that the burst was missing at pin 3 (plate of V702A) the color burst gate stage. Also, the horizontal keying pulse was not found at the grid (pin 2) of this same tube. A few more checks revealed that C712, a ceramic 5000-pf capacitor, had opened, thus accounting for the missing horizontal keying pulse. It is also a good practice to replace the neon bulb (N701) in these chassis, should you have a color reception problem. The color looked good after installing C712 and the neon bulb.

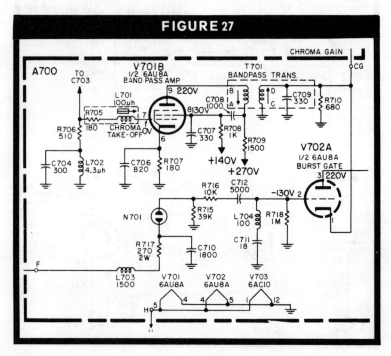

FIGURE 27

GENERAL ELECTRIC CB CHASSIS
Dull, washed-out picture

The brightness control would not extinguish the raster on this set's CRT. However, the contrast control operated a little like the brightness control should. The screen of the CRT was dull, washed out, with retrace lines and very blurred. None of the screen controls had any effect on the picture setup or CRT brightness level.

This problem was caused by the B boost lead wire (the circuit that feeds 1100 volts B boost to one side of the screen controls) being shorted to pin 7, cathode, of the damper tube V104. (Notice the horizontal sweep circuit in Fig. 28.) This was, in essence, putting a high horizontal pulse voltage to the CRT screen controls. This short circuit also burnt up R141, a 100K boost filter resistor, and ruined boost diode CR104.

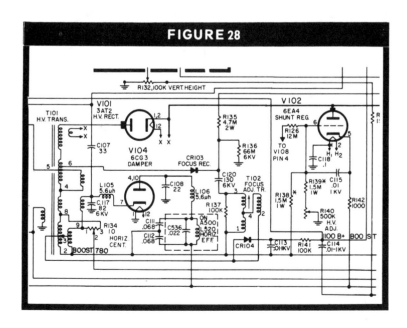

GENERAL ELECTRIC H3 CHASSIS
No raster, no high voltage

Check C266 (.033 mfd) for a short or a leakage condition. (The schematic is shown in Fig. 29.) Also verify that R267 has neither changed in value nor desoldered itself as a result of the heat caused by the C266 short. The resistor may have heated up sufficiently to cause the solder to flow from its leads. Also check V11, the horizontal output tube.

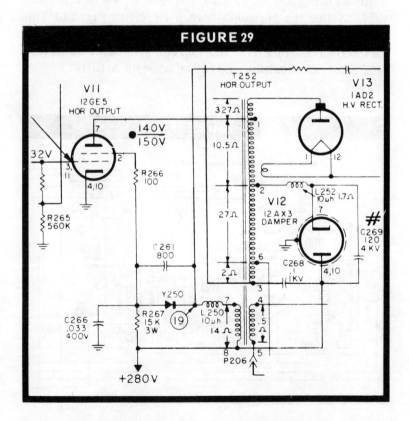

FIGURE 29

GENERAL ELECTRIC H3 CHASSIS
No color or weak color

32

A clue would be low plate voltage at pin 3 of V7A, the color burst gate amplifier shown in the circuit diagram of Fig. 30. If the reading is low, check C512, a .1 mfd 200v Mylar capacitor located at pin 3 of T502. This component has been found to be defective in a number of these chassis. Also, verify that R507 has neither changed in value nor desoldered itself due to the heating.

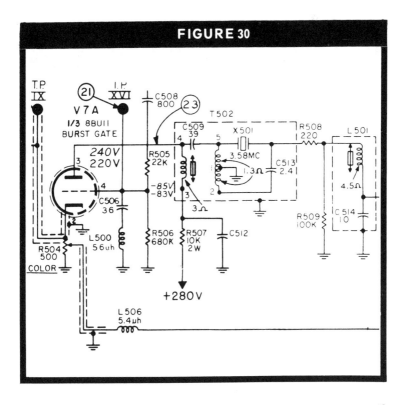

FIGURE 30

33 GENERAL ELECTRIC H3 CHASSIS
Audio distortion

The sound on this color set drifted and became distorted after warmup. To eliminate the trouble, replace C307, shown in Fig. 31, with a GE part number ET18X399 capacitor only. This circuit requires a capacitor with a particular thermal characteristic (N470) to maintain correct alignment. Of course, make the sound alignment adjustments after installing the capacitor.

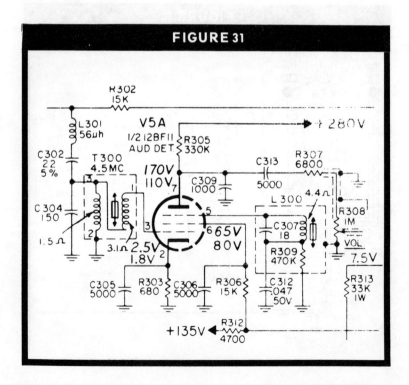

FIGURE 31

GENERAL ELECTRIC H3 CHASSIS
Piecrust (slight) horizontal AFC problem

Circuit Change: Add a 10 pf capacitor between pin 3 of V10 (horizontal oscillator) and chassis common. See Fig. 32. This component may be mounted on the copper pattern side of the printed circuit board of early production version chassis requiring this change. Late production sets have this capacitor mounted on the component side where it is identified as C261.

GENERAL ELECTRIC H3 CHASSIS
No horizontal sweep or high voltage

In this chassis, the horizontal oscillator started intermittently. The cause was found to be C258, a 510 pf capacitor. Use GE part number ET19X86. Refer to Fig. 32 for the horizontal oscillator circuit. Replace C258.

FIGURE 32

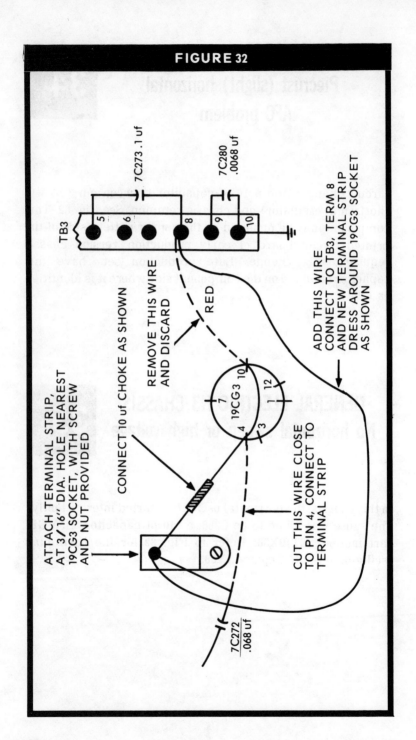

GENERAL ELECTRIC H3 CHASSIS
Vertical jitter

This color chassis had a vertical jitter problem, but all seemed well in the vertical sync and sweep sections and there was no evidence to indicate that other circuits were introducing any interference. The jitter was removed, finally by changing the value of C206 from .033 to .018 and R214 from 12K to 6.8K.

GENERAL ELECTRIC H3 CHASSIS
No vertical sweep

Check C209 (.001 mfd, 20 percent, 2KV) capacitor for a short or leakage. See Fig. 33. If the capacitor is shorted, there will be no sweep at all. If it is leaky, the remaining sweep will be proportional to the amount of leakage.

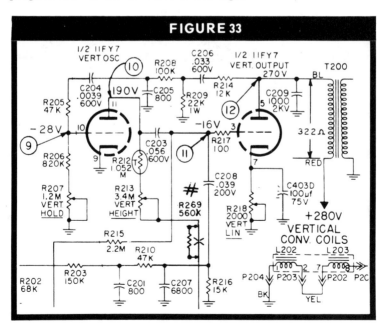

FIGURE 33

GENERAL ELECTRIC KC CHASSIS
Intermittent focus symptom

All circuits seemed to be working normally, but the picture focus was very poor. The screen looked like it had large drops or dots of water all over. You may find this problem on any large screen color receiver that uses a high value resistance in the focus circuit. In some cases, the raster may appear to go-in and out of focus.

Good engineering design requires that a well regulated DC power supply have a constant load at all times in order to have good voltage regulation. A constant voltage is needed to control the focus of the CRT for the sharpest picture. In many color sets this load is a 66-meg resistor designated as R136 in the HV section of the KC chassis shown in Fig. 34. Should this resistor open up or increase in value, the focus voltage will not be constant and thus cause the above condition. Some sets may use a 15-meg and two 22-megohm resistors in series in place of the one 66-megohm resistor.

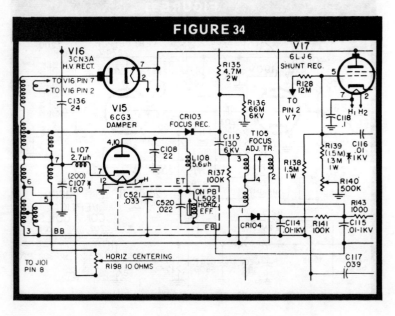

GENERAL ELECTRIC KD CHASSIS
Picture pulls in from sides of screen
39

Insufficient width may occur after a year or so of operation. In most cases, these complaints are not due to the normal component failures.

Trouble is due to a low heater voltage on the 6CG3 damper tube. (In some cases, it checked out at 5 volts.) This low heater voltage is due to a poor contact at the riveted ground or terminal board 2 which carries the heater ground return from pin 12 of the damper.

Adding a length of bus wire from the ground lug of the terminal board to the nearest solder "lance" on the chassis should solve this width problem. Be sure that you get a good soldered joint.

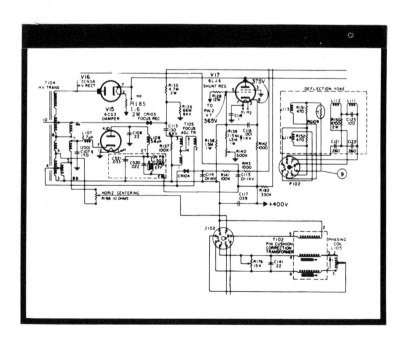

40 GENERAL ELECTRIC KE CHASSIS
Vertical color stripe on the screen

A vertical rainbow stripe of red, green and blue appeared about one-third of the way in from the left-hand side of the screen. The stripe is more visible at low contrast and can occur on both monochrome and color. The horizontal hold control will move the stripe to the left or right.

Probable cause is a fault in the retrace blanking circuit. It may be due to a defective Q203 or some other component in the blanking circuit. Check the voltages at Q203 shown in the schematic of Fig. 35 and use your scope to check for correct waveforms shown at numbers 24 and 25. Notice that the voltage shown at the emitter of Q203 is 2.5 volts.

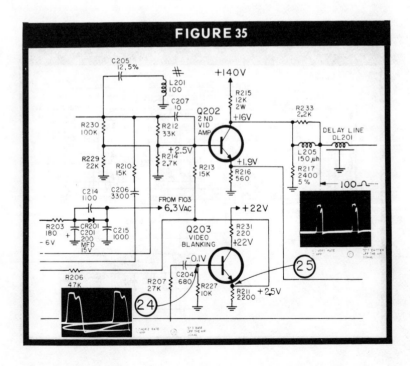

FIGURE 35

GENERAL ELECTRIC KE CHASSIS
Vertical size & retrace lines

41

This GE color set had insufficient vertical picture size, plus vertical retrace lines appeared across the screen. The vertical sweep stages were checked out and no trouble was found.

After a few more checks, the fault was found in T105, the pincushion correction transformer which is shown in Fig. 36. The winding was open between pins 3 and 4. This same trouble has been found in other color chassis.

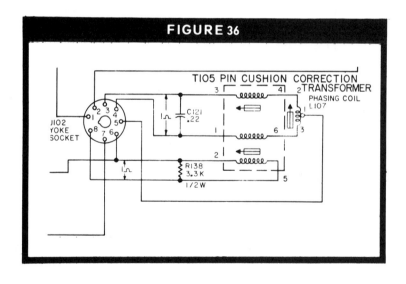

42 GENERAL ELECTRIC KE CHASSIS
Weak video

Weak video was found to be caused by failure of the 6JC6 third IF tube. These tube failures are partly due to defective tubes, but are also due to the following situation: When a strong signal is present as the set is turned on, the tuner and IF amplifiers become operative before the sweep system does because of slow warmup of the horizontal output tube. The AGC keyer stage, therefore, does not bias the tuner and first two IF stages for a few seconds. During this time, the 3rd IF stage, which has 180 volts B+, is overdriven so the grids draw current enough to become overheated until the AGC system begins to operate and reduce the drive to the 3rd IF stage. The result is the loss of Gm in the tube (and chroma) after several months of set operation.

Tip: The tube may short out plate to cathode and overheat plate resistor R315 and R154 and R160 in the power supply. See Fig. 37 for the schematic. Check and replace the resistors if found burnt.

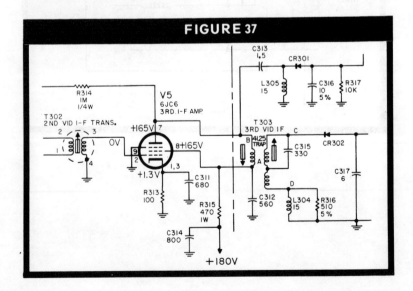

FIGURE 37

GENERAL ELECTRIC KE CHASSIS
Tweet on Channel 5 or 11

43

In some reception areas, you may encounter an interference problem which produces a beat pattern as shown in Fig. 38. It is usually found when using "Rabbit Ears" or when an unbalanced condition exists in the antenna or CATV system.

This interference or tweet may be found on either Channel 5 or 11, but will occur only in the presence of burst in the signal. Therefore, it will appear during a regular color transmission or a monochrome transmission if burst is present (as is the case for some commercials). Also, turning the tint control will change the speed, intensity or pattern of the beat.

Correction: Add a 20 pf Hi Q ceramic capacitor on the bottom of the main circuit board as follows: Using the shortest possible leads, connect the capacitor from the chroma gain control end of L725 at the wire wrap terminal to the nearest solder point on the copper ground plane.

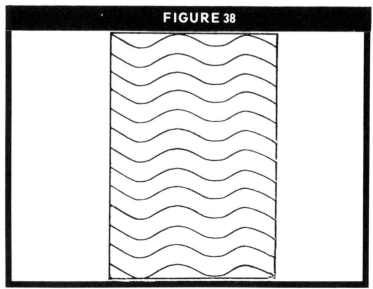

FIGURE 38

44 GENERAL ELECTRIC KE CHASSIS
Raster Shading Problems

A raster blanked out at normal to high brightness settings may be caused by an open coil (L203) in plate circuit of the video output stage, just ahead of the delay line. (For this and the following problems, refer to Fig. 39.)

A floating gray horizontal bar across the screen, with some shading to top of screen, perhaps, may be the result of an open filter capacitor, C201 or C202, 200 mfd at 15v DC.

An open capacitor C152-B, located in the power supply section, will cause a floating horizontal black bar. Light shading on the left side of the raster, while the right side is very dark, could be due to an open C203, a 6800-pf capacitor in the emitter circuit of Q201 (1st video amplifier).

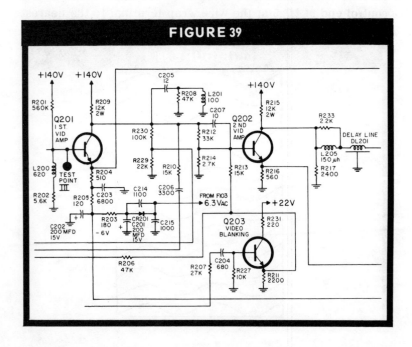

FIGURE 39

GENERAL ELECTRIC KE CHASSIS
Horizontal Pull

45

The complaint for this chassis was a type of horizontal pulling. Right off, the horizontal sync and AFC phase detector section was checked, but no circuit faults were found. This can be a very tough symptom to isolate. The scope found the B+ filtering system to be in good condition. However, the trouble did not look like an AGC trouble, but it was. After a lot of troubleshooting, the trouble was found to be a defective AGC control. See Fig. 40. After changing R454 and adjusting the AGC level, the picture locked in rock solid.

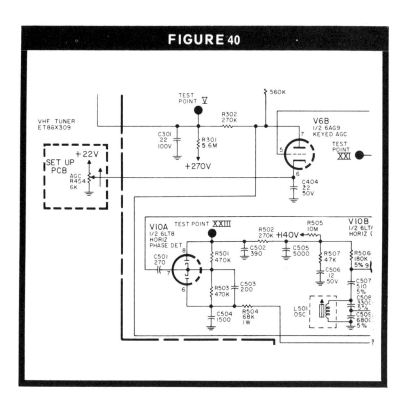

GENERAL ELECTRIC KE CHASSIS
Picture bending or snow

A faulty capacitor may cause double-trouble in some of these GE KE receivers. Refer to the schematic in Fig. 41.

One of these sets came into the shop with low sensitivity and had a very snowy picture. The AGC system was checked and capacitor C305 (a .1 mfd, 100 volt value) was found to be shorted. This is the AGC filtering capacitor.

Another chassis displayed a bending-pulling picture and had a very loose sync-lock condition. And again, C305 was found to be defective, but this one was now open. The last condition seems to be rare.

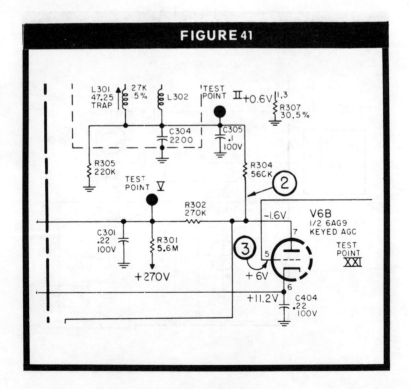

FIGURE 41

GENERAL ELECTRIC KE CHASSIS
Video troubles

47

The causes of the following video troubles appear on the schematic in Fig. 42:

No video or very weak video. Capacitor C203 has shorted.

No video. Coil L308 has opened up.

No video or weak, washed-out video. Transistor Q202 is defective.

No video or audio and possibly some vertical retrace lines. The 200 mfd filter capacitor, C202, has shorted.

Blank screen, no video. Faulty Q201 transistor.

FIGURE 42

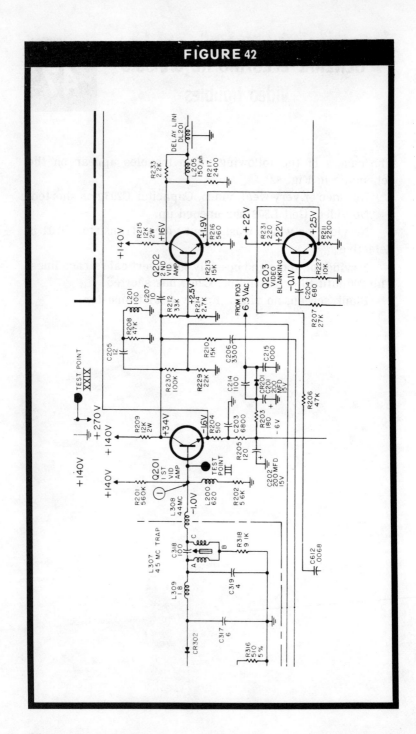

GENERAL ELECTRIC KE CHASSIS
Raster pulled in from sides

The raster on this receiver was pulled in from the sides. The grid drive to the horizontal output tube was low due to a defective capacitor (C512) in the horizontal oscillator section. See Fig. 43.

GENERAL ELECTRIC KE CHASSIS
Lines across screen

The horizontal oscillator was found to be off frequency in this color set. If the oscillator is far enough off frequency the horizontal output tube will overheat and no high-voltage would be developed. The trouble was found to be a faulty 510 pf capacitor (C507) in the horizontal oscillator circuit shown in Fig. 43.

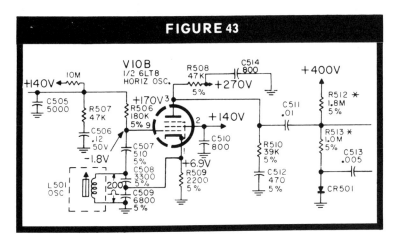

FIGURE 43

MAGNAVOX
Failure of the color detector diodes

Use this service tip if you find a defective diode in either the color killer or color phase detector circuits. In some cases, arcing in the picture tube may be causing these diode failures. Spark gaps are provided on the chroma board for the CRT grid leads. In the case of the blue grid (term. DD) the spark gap ground is connected at the ground land for V706. The spark gap should be re-located so that it returns to a more direct path to chassis ground. In current production it is being connected to the ground lug on the terminal strip used to mount R147 and R146 (in the Chromatone circuit). Make this change if you have repeated diode failures.

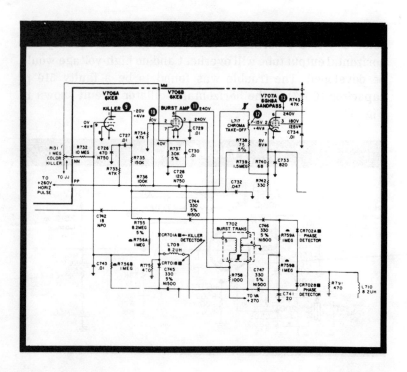

MAGNAVOX T924 CHASSIS
Picture top hook

51

This color chassis displayed a picture with a hook up at the top. This symptom did not change from scene to scene and no binding or weaving was noted. The sync and B+ voltage supply section checked out good and there was no visible hum seen in the picture. In fact, all of the circuits appeared to be functioning properly.

To straighten out this picture hook, a circuit modification is needed. Refer to the sync separator circuit schematic in Fig. 44. Add a 1K ½-watt resistor in series with a .1-mfd capacitor from point 7P on the chroma board (cathode of V703A) to ground. The resistor is to be connected to point 7P and the capacitor goes to ground. Also, remove C125 (.01 mfd, 500-volt) from the circuit. The capacitor, not shown, connects to the AGC control.

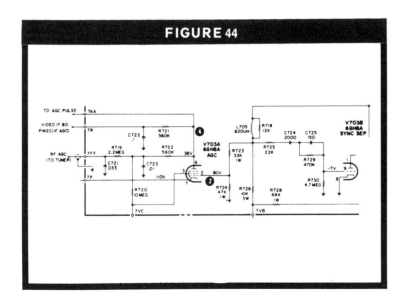

FIGURE 44

52 MAGNAVOX T933 CHASSIS
Critical horizontal locking action

Some of these chassis have developed a very critical horizontal hold. This horizontal "lock out" symptom occurs as the receiver is switched from one channel to another. The only way proper horizontal picture lock can again be restored is by re-adjusting the horizontal hold control.

To correct this condition, make the following circuit change. Refer to the sync separator stage, V703B, shown in Fig. 45. Install a 12 meg resistor, ½ watt, from pin 1 to pin 9 of V703, the sync separator stage. You should now have solid horizontal locking action.

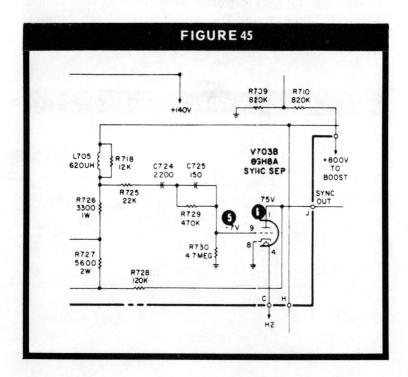

MAGNAVOX T933 CHASSIS
Critical color sync

This set had critical color sync, depending on the setting of the hue control, and adjusting the reactance coil would cause it to lose sync and color completely. Tubes and diodes checked good. The cathode voltage on the burst amplifier (V706B) was the key measurement, as it was about 15v DC instead of the usual 40-50v as shown in the schematic of Fig. 46. The scope showed a severe distortion of the burst signal at the plate and an amplitude of 7v p-p instead of the normal 100v p-p. The fault was an open .01 mfd capacitor (C730) in the burst amplifier circuit. With this capacitor open, degeneration in the cathode circuit made the burst amplifier ineffective and there was not enough burst signal to provide any color sync.

MAGNAVOX T933 CHASSIS
Overall tinted raster and low color saturation

Another T933 chassis had what looked like phase detector troubles. The entire screen had a tint which could be varied (blue, magenta to red) by adjusting the hue control and could be eliminated by turning the chroma control down. This condition only showed up when a color program was being received. The set was checked with a color bar generator and the bars on the right side of the screen showed little of the correct colors. Instead, they assumed the overall raster tint. This test proved that the color signal which did get through was in color sync and the hue control could vary the colors of the bars. Using a scope to trace this color signal through the chroma circuits revealed a rather husky horizontal pulse on the grid of the bandpass amplifier, V707A. Again refer to Fig. 46. The plate of the color killer stage is coupled to the grid of the bandpass amplifier to cut off this stage when a B&W signal is received. The color killer uses a horizontal pulse as its plate voltage source (like a keyed AGC stage) but there is a .047 mfd capacitor (C732) in the bandpass grid that will by-pass to ground any horizontal pulse present there. In this case, C732 was found to be open.

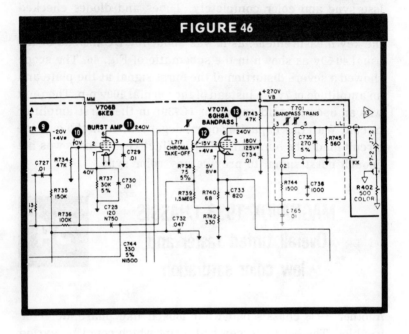

FIGURE 46

MAGNAVOX T935 CHASSIS
Vertical jitter

55

Use this service tip to eliminate vertical jitter in this Magnavox receiver. The jitter problem in this chassis can be corrected by removing capacitor C54 (4 mfd, 25v), located on the horizontal circuit board, and reinstalling it on the VHF tuner directly between the RF AGC terminal and ground. Install a 4 mfd, 24v capacitor on the VHF tuner directly between the +18v DC terminal and ground. Install a 1000 mh choke, part number 361324-102, in series with the +18v DC terminal of the VHF tuner. These changes will eliminate the vertical jitter condition.

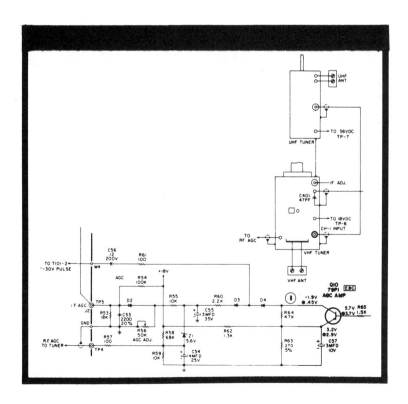

MAGNAVOX T938 CHASSIS
Weak color

This set had a good black-and-white picture, but the color was weak. No trouble was found in the antenna system and all tubes checked out good. On the service bench the scope was used to trace the chroma signal through the color amplifier stages. The scope found a very low p-p chroma signal at pin 6, plate of V707A, color bandpass amplifier, shown in the schematic of Fig. 47. A few more checks in this circuit revealed that screen bypass capacitor C734 had shorted and caused R743 to burn. We replaced C734, a .01 mfd, and R743, a 47K resistor, and the set had good strong color.

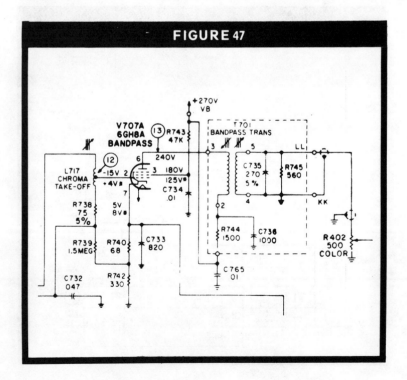

FIGURE 47

MAGNAVOX T908-T915 CHASSIS
Unstable picture

57

In some extremely noisy signal areas, both sync and AGC may be affected. Make the following changes to improve noise immunity.

1. Change C233, a 50 mfd capacitor in the emitter circuit of the AGC keyer to .15 mfd.

2. Add a 4 mfd capacitor from the junction of diode D204 and R235 to chassis ground.

3. Change R241 from 100K to 470K and connect from the base of the sync separator (Q208) to point 2D in the AGC circuit. R241 was connected from the base of Q208 to +12 volts.

4. Change capacitor C403 from .033 to .002 mfd and C402 from .02 to .01 mfd.

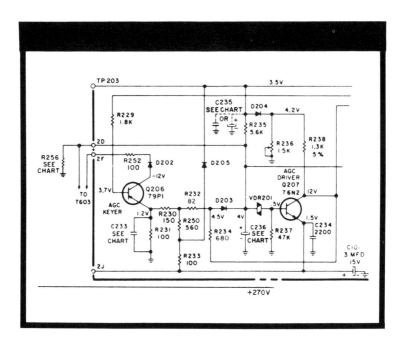

58 MAGNAVOX T919 CHASSIS
Low brightness

Reports of low brightness indicate the +140v source may be too low in some receivers. This +140v supply is used in developing proper bias for the 12GN7 video output stage, and the DC plate voltage of this stage sets the bias for the CRT cathodes. A low 140-volt source, therefore, will reduce the output of the 12GN7 and result in a reduction of brightness. To bring the +140v up, connect a 3.6K 7w resistor across R102 (1100 ohms, 18w) and a 5.6K 5w resistor across R160 (910 ohms, 7w) in the power supply section.

MAGNAVOX T936 CHASSIS
Weak video

59

The weak video showing on the screen of this set looked like we had a weak CRT. However, the picture tube checked out very good on this small screen color TV receiver. The DC voltages on the 9KX6 (shown in the video output schematic of Fig. 48) were measured and found to be very close to normal. An oscilloscope was used to trace the video signal through all of the video amplifier stages. We found the proper video amplitude signal at the grid (pin 2) of V107, but a very low amplitude on the plate (pin 7). The trouble was found to be an open 100 mfd capacitor in the cathode circuit of V107. After C141 was replaced, the set produced a good picture with lots of contrast.

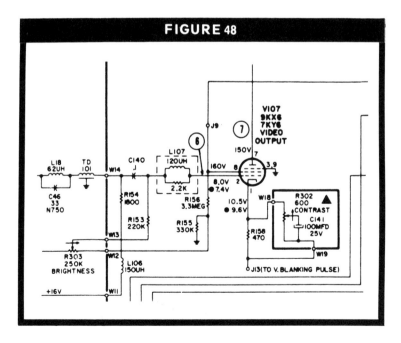

MAGNAVOX T931, T933 AND T940 CHASSIS
No color or poor color sync

Intermittent loss of color sync, poor color sync or loss of color may be caused by the diodes in the color-killer detector or in the color-phase detector. These diodes are to be matched pairs. If one of the diodes should undergo a change of characteristics, circuit operation may be impaired. Use the following test for any suspected diode:

Connect a color-bar generator to the receiver and obtain a normal color-bar pattern display. Remove the 3.58 MHz oscillator tube (V708) and ground the junction of R756A and R756B in the color killer detector circuit. Use a VTVM and measure the DC voltage to ground from the outside end of each diode pair. The voltage from ground to the cathode will be positive and negative from ground to the anode. The exact value of this voltage will vary from set to set and with the level of the color burst signal. (Typical values may be -55 volts at the anode and +55 volts at the cathode.) The service "tip" is the difference, if any, in the voltages measured across each diode pair. Ideally, the voltages across a diode pair will be identical, but a 10 percent variation is allowable. If a voltage difference of more than 10 percent is noted across either diode pair, replace that pair with a matched pair, part number 170733-1. Do not replace just one diode of a pair.

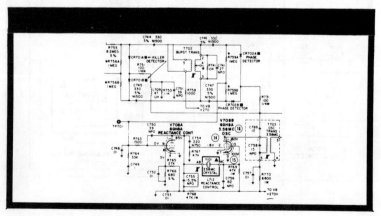

MAGNAVOX
Remote control receiver model 704058 station skipping. May be intermittent

61

Adjustment of the horizontal hold, remote sensitivity and HV are all important to proper operation of the remote receiver and in some cases may cause channel skipping.

The service tips are as follows:

1. Check the horizontal hold to insure that the horizontal sync locks in for all stations. Note: If the horizontal oscillator is running off frequency due to a change in value of resistor R533, this can cause channel skipping. Check R533 and replace if off value.

2. Check the setting of the search sensitivity control. If set too low, this will cause skip.

3. Check the high voltage and adjust the HV control as necessary to provide 24.5 KV.

On these color chassis the collector of transistor Q1, the coincidence gate transistor located on the AFT board of the TV chassis, and the base circuit of Q20, the sync gate driver transistor located on the remote receiver chassis, obtain their operating voltages from a 60v pulse supplied from the horizontal output transformer. When the tuner is off-channel, Q1 will be cut off because sync pulses are not present to bias it into conduction. When a station is tuned in, the sync pulses provide the saturation bias for Q1. When Q1 saturates, the forward bias voltage to Q20 is reduced to below cut-off.

If the HV adjustment is set to increase the high voltage appreciably above its normal value, the loading on the horizontal sweep circuit will be reduced and the amplitude of the pulse voltages in this sweep circuit will increase. This increase in pulse voltage can be sufficient to keep the forward bias of Q20 from being reduced to cut-off value, resulting in the remote receiver search circuit not recognizing that a station has been located.

If you encounter a problem of "station skipping" and all other conditions, including the adjustments of the AGC, horizontal hold and remote sensitivity controls, seem normal, check the value of the HV and readjust as required.

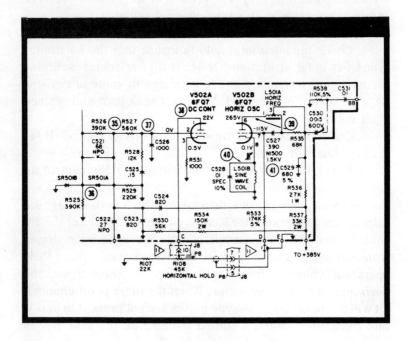

MOTOROLA TS-908 CHASSIS
Vertical white drive line

62

This set produced a good picture except for a vertical white drive line near the center of the screen only during B&W reception.

To cure this problem, make the following checks and changes: Tighten the screws that clamp the horizontal output transformer cores together. Try a new horizontal oscillator tube (V13) and horizontal output tubes. Change resistor R512 at the plate of V13B from 22K to 27K. Also, resistor R514 can be changed to 820K or 680K, provided this does not reduce the grid bias of each horizontal output tube to a lower value of voltage than shown in Fig. 49. On some sets this drive line is usually slight and observable on unused channels only.

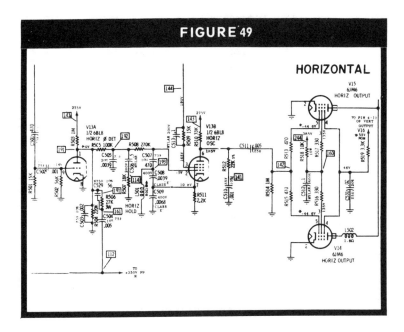

MOTOROLA TS-908 CHASSIS
Picture side pull

The picture was shrinking in from both sides of this color chassis. This could be trouble in the horizontal sweep output section or even in the power supply. The B+ voltage was measured and found to be correct. The cathode, screen, and control grid voltages were normal for the horizontal output tube. Other voltages in the horizontal output section were checked and found to be within tolerance. Note the schematic for the horizontal sweep circuits in Fig. 50. It was also noted that the high voltage would be lost intermittently. It was found that these picture problems were caused by shorted turns in the horizontal linearity coil, or the horizontal efficiency coil. Generally this fault can be detected by noticing if the coils are burnt or are operating at a very high temperature. Check and replace defective component coils L504 and L508.

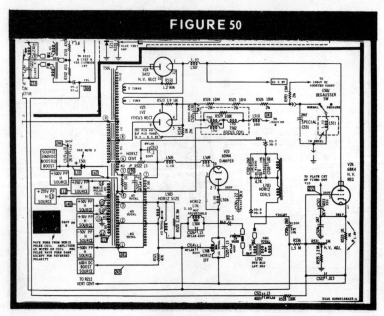

FIGURE 50

MOTOROLA TS-908 CHASSIS
Low picture brightness

64

This color receiver produced good audio; however, the screen had insufficient brightness. The CRT and the high voltage checked out good. Some checks were made in the video stages, but no circuit faults were found. The defect was located in the blanking amplifier circuit shown in Fig. 51. The plate resistor of V20B was found to be open. This open resistor, R920 (39K, 2 watt), lowered the plate voltage on pin 1 of the 6BL6 blanker tube. This same symptom can be caused by an open C917, .1 mfd capacitor not shown in the schematics, also, in the plate circuit of V20B. Don't overlook a faulty (weak) 6BL6 blanker tube.

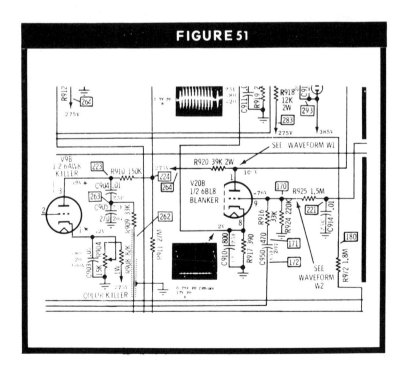

FIGURE 51

MOTOROLA TS-914 CHASSIS
Circuit modification for insufficient brightness range

On early run chassis, it was possible to adjust the brightness so high as to exceed the power handling ability of the HV rectifier tube (3AT2). Beginning with TS914A-05 chassis run, the brightness range was restricted by removing B+ (385 volts) from the top end of the master G1 control and connecting a 47K resistor from control arm to ground. On these chassis, if more brightness is required, first try a new 6LY8 video tube. If this does not help, resistor R224 can be changed from 47K to 150K. Later production chassis now use 150K resistor as shown in Fig. 52.

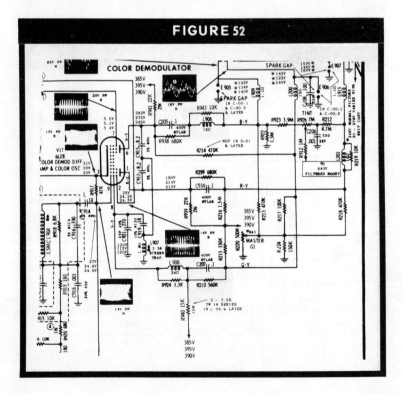

FIGURE 52

MOTOROLA TS-914 CHASSIS
No video information

This set had good sound but no picture, just a blank white screen. The HV and brightness were normal. Voltage checks were not of any help for this trouble. The fault for this symptom was located in the video IF section that is shown in Fig. 53. This (no video) condition is caused by an open 200-ohm control (R115) located in the T102 transformer circuit. For a quick check, just jump a test lead across the control. Replace R115 control and check the IF alignment.

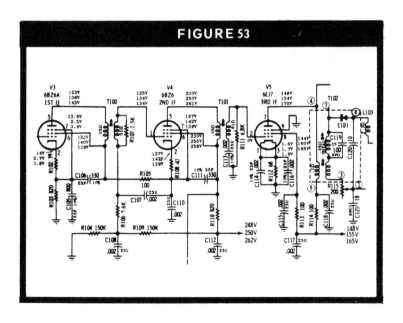

FIGURE 53

67 MOTOROLA TS 914-918
Picture not clear and poor peripheral convergence

The convergence board was checked and found to be in good health. The next step was to use the oscilloscope to test for correct convergence pulses. The scope told us there was no horizontal pulse present at the convergence board. The trouble was located as an open pulse winding on the horizontal sweep transformer shown in the circuit diagram of Fig. 54. The pulse winding was replaced and picture convergence was now good. The Motorola part number for this coil is 24D66875A10.

68 MOTOROLA TS 914-918
Screen will not light up and no HV at the anode cup

The filaments on HV rectifier would not light, but lots of RF was present at the top cap. A check revealed that the pulse winding had shorted, acting as shorted turn on HV sweep transformer, killing the filament voltage. See Fig. 54. Replace the pulse winding (see Part No. above) of the HV transformer, and check and adjust the HV.

FIGURE 54

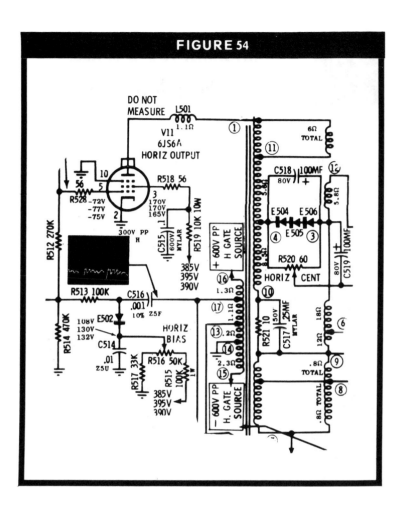

69 MOTOROLA TS-918 CHASSIS
Color killer critical to adjust

Color circuit modification. Refer to Fig. 55. If you come across a chassis where it is difficult to adjust the color killer control to eliminate color stage amplification on B&W signals and still allow normal color operation, change capacitor C923 in the plate of the killer tube from 220 pf to 120 pf. Also change R940 (not shown in the schematic) in the screen of the demodulator tube from 10K to 15K and connect the bottom end of this resistor to 385 volts B+. These changes have been made in later production chassis.

70 MOTOROLA TS-918 CHASSIS
No color

Check color IF transformer, T900, for an open primary or shorted secondary winding. Again, refer to Fig. 55. The defect in this set was caused by a capacitor lead which is connected to the secondary and is dressed across the primary. Later production sets have an insulating sleeve added to the capacitor lead to prevent this type of short.

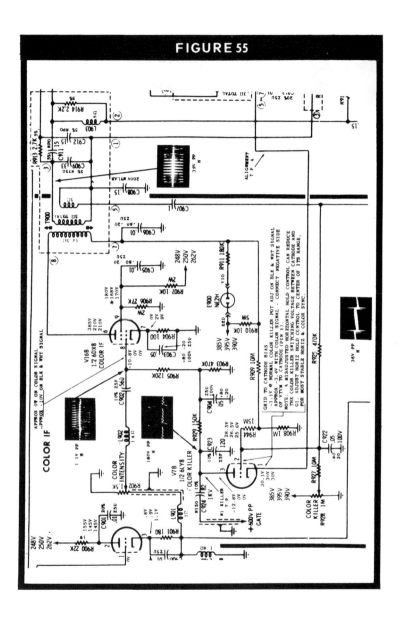

FIGURE 55

87

MOTOROLA TS-918 CHASSIS
No sound

This set displayed a good picture, but there was no sound to be heard. With the volume control up to its maximum setting we could detect a slight arcing sound with the ear close to the speaker. A few voltage checks in the audio section, see Fig. 56, isolated the problem to the output audio stage, V9B. Only about 30 volts was measured at the plate of V9B. Capacitor C317 had developed leakage (or may become shorted) and had lowered the plate voltage. Replace C317 with a .001 mfd, 2 KV capacitor. Also, check audio transformer T301 for shorted windings and possible internal arcing.

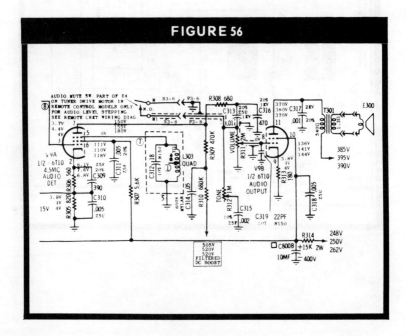

FIGURE 56

MOTOROLA TS-918 CHASSIS
Focus problem

72

The picture was not clear on this color set and the focus would appear to change erratically. The trouble, in this case, was a defective focus divider resistor, R524, which is located in the HV cage. (See the schematic in Fig. 57.) This resistor runs hot and sometimes arcs or becomes burnt. This focus condition will vary by the length of time the set is playing. Usually, the longer it operates the more erratic this focus change becomes. Replace the R524 focus resistor, check R526, the 10 megohm focus control, and also V13, HV rectifier tube. Check the high voltage and adjust the focus control for a sharp picture.

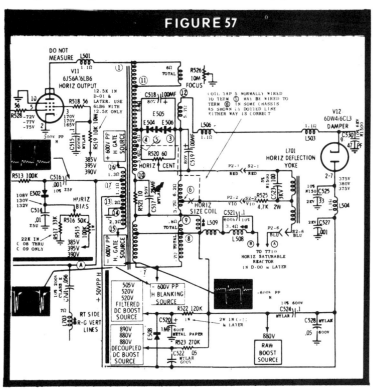

MOTOROLA TS-918 CHASSIS
Gray vertical bars

The picture on this receiver had vertical gray bars all the way across the screen. The color picture was also missing. This is a very "odd ball" trouble and can be very tough to locate if you don't know what to look for. The voltages all checked out and a visual inspection did not reveal any faults. In this case the oscilloscope gave some erroneous indications. These gray bars looked like a color-bar generator pattern, but without the color. This bar pattern was caused by a defective 3.58 MHz crystal, E901, shown in the schematic Fig. 58 of the color sync circuits. This same problem has appeared in other brands of color TV receivers.

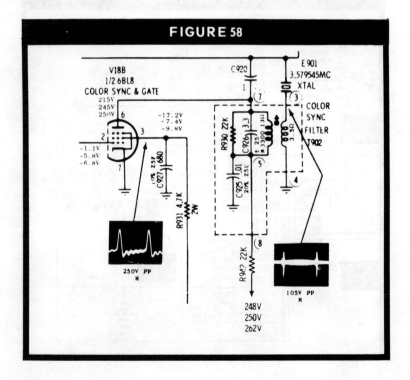

MOTOROLA TS-918 CHASSIS
Low brightness level

74

This color receiver would operate with a normal picture for about ten minutes and then would dim down to a low brightness level. In many cases this will be an intermittent picture brightness condition. Many defective components or faulty circuit connections can cause this same symptom. A good place to look for this type trouble is in the video amplifier stages. And if you suspect a thermal type of malfunction, use some circuit cooler spray to isolate the defect. This solved the problem fast for this chassis. This set had an intermittently open peaking coil, L105, shown in Fig. 59. The voltage at pin 6, plate of the 60X8 tube, measures low when the coil opens. The low brightness problem was solved when L105 was replaced.

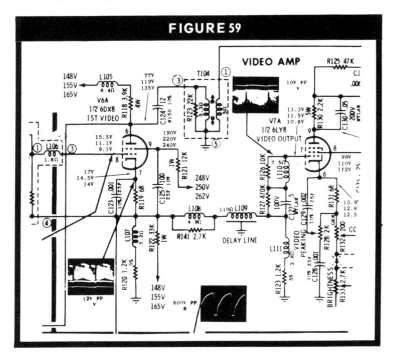

FIGURE 59

75 MOTOROLA TS-918 CHASSIS
Blurred video or low brightness

The first problem for this chassis was a picture with blurred video and a white vertical bar on the screen that looked like some type of horizontal foldover. The picture would bloom when the brightness control was turned up. This symptom was caused by an open choke coil, L113, shown in Fig. 60 (video output stage schematic). Replace this coil, which is mounted near the video drive controls.

The second problem this chassis developed was low brightness on the CRT, even with all the controls set at maximum. The high voltage at the CRT anode was also correct. This low brightness condition was caused when R131 in the cathode circuit of V7A increased in value. Replace R131 with a 68 ohm resistor and readjust the video drive and CRT bias controls.

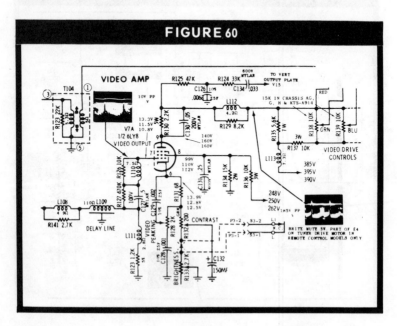

FIGURE 60

MOTOROLA TS-915-919 CHASSIS
Picture blue and not clear

76

This solid-state chasis had lots of blue bleeding over the entire screen and was not producing a very clear picture. A crosshatch generator pattern was displayed on the screen and we could see the bleeding was still there. A touch-up of the static and dynamic convergence controls did not help at all. However, all of the controls appeared to be working. Of course, the convergence "H" panel was a prime suspect and a new one was snapped into place, but this did not solve the problem. The panel circuit schematic appears in Fig. 61.

It was now time to use the oscilloscope and check the pulses going to the convergence panel. We looked at the parabolic current coming out of this panel and being fed to the convergence coils around the neck of the CRT. All of these waveshapes were found to be correct. The trouble must be in the convergence coils and a finger check of the blue one uncovered the trouble. It was very warm to the touch.

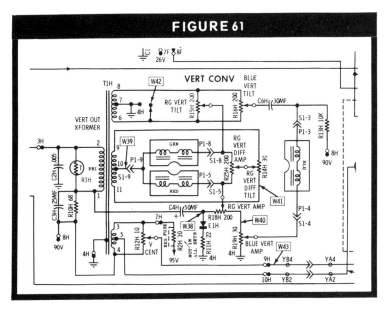

77 MOTOROLA TS-915-919 CHASSIS
Narrow picture

In this solid-state chassis we had good sound and a picture, but it was very narrow. The first section to be checked was the power supply and then the horizontal sweep circuits. However, these systems were all found to be in good order. After looking at Motorola's trouble chart, we checked pincushion panel "G." This panel is shown in the Fig. 62 schematic. This narrow picture was caused by a defective 10 mfd capacitor located in the emitter circuit of Q3G, the horizontal regulator amplifier transistor. Replace C6G.

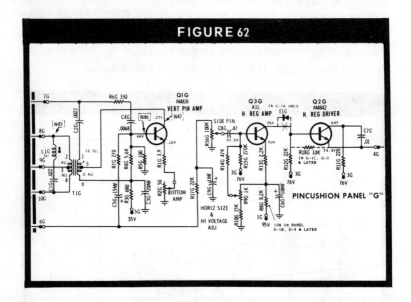

FIGURE 62

MOTOROLA TS-915-919 CHASSIS
No high voltage

78

This set did not produce and no HV was found at the CRT anode lead connection. No trouble could be found in the horizontal sweep and HV section. A complete check of the power supply and low-voltage system revealed no faults. After a new pincushion panel "G" was installed, the picture and HV were restored. We then started troubleshooting the "G" panel. The fault was an open resistor in the emitter circuit of Q2G, the horizontal regulator driver transistor. Replace R11G, a 220-ohm resistor shown in Fig. 63.

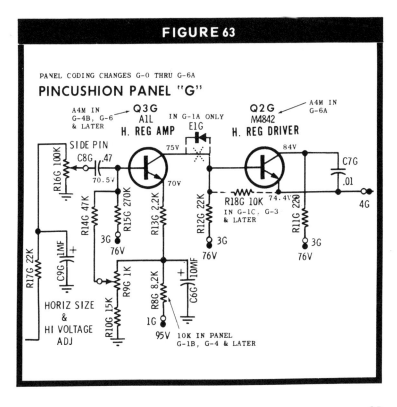

FIGURE 63

79 MOTOROLA TS-924 CHASSIS
Poor resolution

This receiver had good sound, but a poor focus condition. At least that's what it appeared to be. A closer observation revealed that the raster had sharp focus, but the degraded video information displayed on the screen was due to poor resolution.

This poor resolution problem may be caused by several different circuit faults. Use these following suggestions for a check list. See the video circuit in Fig. 64 as you go through this list.

1. Check for a too-narrow bandwidth in the IF amplifier section.

2. The detector load resistor may have increased in value.

3. Check for a faulty R31B resistor (not shown) video peaking coils, and delay line, L128.

4. Check all voltages on the V6 and V7.

5. Measure all video amplifier tube (V6 and V7) load resistors.

6. Of course, make a complete IF sweep alignment check.

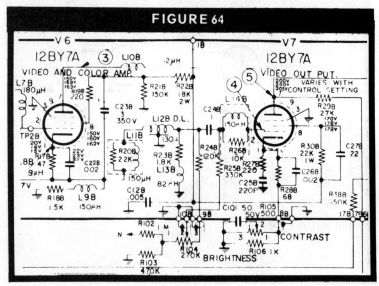

FIGURE 64

MOTOROLA TS-934 A CHASSIS
Hum bar

80

Many of these Motorola Quasar II Color Chassis develop tne same picture symptom, but more than one fault will cause this trouble. The picture would show a horizontal hum bar across the screen (light shaded bar that may or may not move up across the screen. And, of course, it was of an intermittent nature. On some sets a slight bump on the chassis would clear up the hum for a short time.

The first trouble was located on the ZA power supply (see Fig. 65) and DC regulator panel. It was found that the large C3 capacitor (a 1000 mfd unit) would break loose on the circuit board and required resoldering. Also some of these ZA panels were not making a good chassis ground connection. To insure a good connection, scrape the insulation coating from around the ground mounting clips and make sure the panel fits tight to insure a good ground connection. Also check and make sure capacitor C2 has a good electrical solder connection to the panel.

Another fault in the main chassis frame will also cause this same hum bar problem. And this is C205, a 10 mfd capacitor that is part of the power supply filtering system. This capacitor has been known to open up. Should you have a hum bar problem in this chassis the above "tips" will clear the picture up considerably.

FIGURE 65

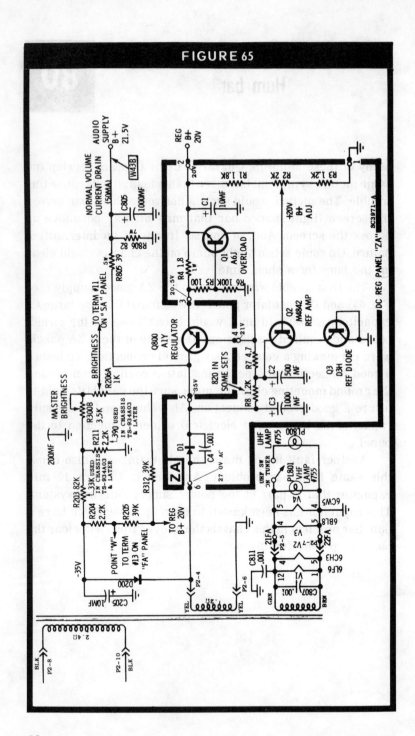

PHILCO 19 FT20 AND OTHER CHASSIS
Vertical intensity ringing (bars on the left side of screen)

In the event you find vertical bars appearing on the left side of the screen, check for proper dress of the purple lead going from pin 8 of the convergence socket (J201) to lug M47 on the chroma panel. This lead should be dressed as far away as possible from the video IF panel toward the front apron and down as close as possible to the chassis sub-base.

For chassis using the 21-inch round picture tube, some arcing may show up as vertical line interference patterns on the screen.

Probable Cause: The 1K resistor in series with the high voltage lead is defective (open) or the plate lead and socket of the 6BK4 tube is faulty. Replace the socket and lead assembly with Philco part number 41-4367-1.

82 PHILCO 19FT20 CHASSIS
Vertical buzz in the sound

For this color chassis use the following tips, should you encounter some vertical buzz in the audio.

If the buzz is electrical only, that is, cannot be heard with speakers disconnected, check for the proper lead dress from the plate of the vertical output tube to the vertical output transformer under the chassis. This lead must not cross over the audio output socket. If the buzz is mechanical and can be heard with the speakers disconnected, replace the vertical output transformer. In some isolated cases it may be necessary to move the audio output transformer to the top of the chassis or rotate its mounting about 90 degrees to reduce the magnetic induction directly from the vertical output transformer to the audio output transformer. This would be a rare problem.

PHILCO 19FT20 CHASSIS
Picture lacks detail

83

Color picture lacks detail or the picture appears out of focus, but the focus voltage is correct. Check and replace delay line DL101 if it is shorted or has leakage to ground. If the line is shorted, the signal will not be delayed and will lead the color signal on the CRT.

If the color picture lacks detail (appears out of focus) and is getting no monochrome signal, check and replace the delay line if found to be open. Without Y, the color signal has no detail. Note video circuit in Fig. 66.

Delay lines can have three possible faults.

1. Coils that are shorted to ground.
2. An open coil circuit.
3. Internal capacitors to ground may be open. A common trouble is an open coil and some of the older color chassis delay lines were open on the ground side. This would cause standing waves and produce ghosts on the picture. These ghosts could not be eliminated by antenna adjustment or by the fine tuning control. It should be noted that misalignment in the IF stages will cause ringing and this does change with fine tuning adjustments. If the delay line is suspected, clip a jumper test lead across the two video signal leads. If the ghosts get lost and the picture clears up, this proves the delay line is defective. Also, a short from the delay line to ground, or an open delay line, will, in most sets, cause a loss of both raster and picture information.

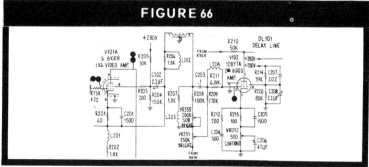

FIGURE 66

84
PHILCO 17KT50 CHASSIS
One color missing

See Fig. 67. Check the appropriate demodulator, X for red and Z for blue. On some models the R-Y amplifier and the X demodulator use the same tube. The Z demodulator and the B-Y amplifier also use a common tube. The G-Y amplifier is common with the second chroma amplifier. A defective tube, in this instance, can affect green or all three colors. Check the matched pair of plate resistors (R70, R82) in the X and Z demodulators. Check the color drive pots in the CRT cathode circuits by checking the cathode voltage on the appropriate CRT gun (M58, M59, and M60). Check the CRT screen voltages at M120, M121, and M122 as the case may be. And don't overlook the color picture tube as being defective.

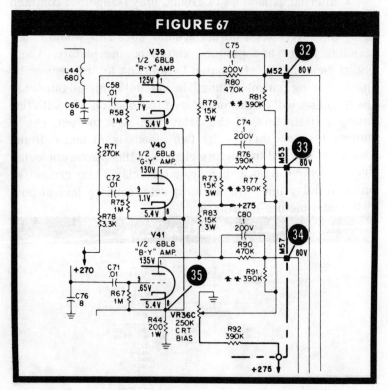

FIGURE 67

PHILCO 17KT50 CHASSIS
Vertical jitter

On some of these chassis we have seen cases where capacitor C117 in the vertical oscillator cathode was a .01 mfd 200-volt. This should be a .01-mfd rated at a 600 volts. In the event of vertical jitter on the bottom of the raster or foldover, C117 should be checked to determine if the voltage rating is correct. If not, make sure a 600-volt rated capacitor is used. See schematic in Fig. 68. If you have one of these chassis in the shop for other repairs, C117 should be checked for proper voltage rating, and if incorrect, it should be replaced to prevent failure in the future.

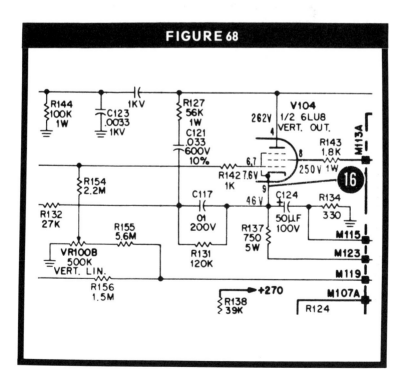

FIGURE 68

86 PHILCO 20KT40 CHASSIS
Color lacking red

The color chassis had a good monochrome picture, but the reds were missing. We made a few fast checks with the VTVM and all of the voltages were normal in the chroma stages. This looked like trouble in the demodulators, so the scope was used to find the defective component.

The missing red was caused by a defective transformer, L103; see the schematic in Fig. 69. The secondary of L103 had opened up between pins 2 and 6.

Another 20KT40 chassis had an odd color symptom. This set had color only on the left side of the raster when the chroma control level was turned up to a maximum setting. This half-color condition was caused by a defective Q102, the 2nd chroma transistor amplifier stage.

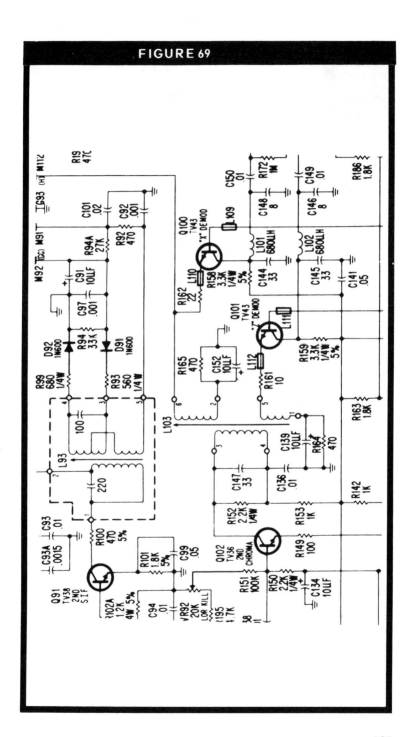

FIGURE 69

87   PHILCO 20KT40B CHASSIS
Critical horizontal sync

The horizontal would "sync-in" on this set, but wouldn't lock with very much stability. When the channels were changed, the picture would go off horizontally. The horizontal AFC system was a prime suspect; however, a few checks cleared these circuits. The defect was found in the horizontal reactance control circuit. This critical horizontal sync condition was caused by a faulty (leaky) capacitor C64. This is a 1-mfd, 15v unit located in the cathode circuit of V42, shown in the horizontal control circuit schematic of Fig. 70. Replace capacitor C64, check the 6BL8 tube and adjust the horizontal hold control.

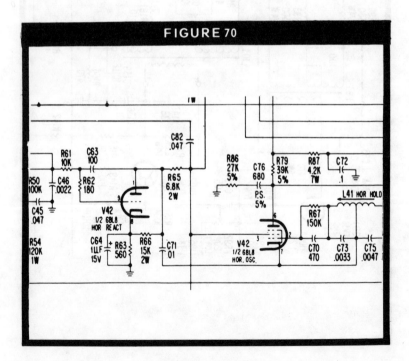

FIGURE 70

PHILCO 20KT41B CHASSIS
Loss of color

This chassis produced a B&W picture that also had some intermittent color. We also noted that the blue at the top of the raster could not be converged properly. After many routine tube and transistor checks, the color would still fade out. As you well know, this type of intermittent problem can be very exasperating. The oscilloscope was put into action in the color section and the horizontal keying pulse was found to be very low in amplitude. Of course, the "tip-off" to this problem was the fact that the chroma and convergence circuits were not operating correctly. This trouble was caused by a poor ground lead connection at the horizontal pulse winding ("Y") of the horizontal sweep transformer. Note the schematic in Fig. 71. This winding (red and white lead) feeds a horizontal pulse to the convergence and chroma panel circuit boards. The ground side of the winding may not be making a good connection to the horizontal output transformer bracket. Check and resolder this ground lead.

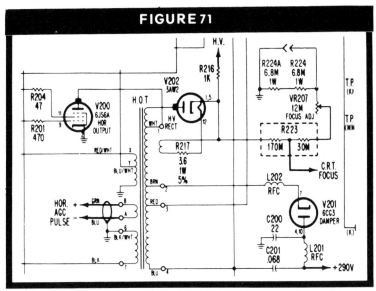

FIGURE 71

89
PHILCO 12L80 AND 13L80 COLOR CHASSIS
Poor sync

Poor vertical sync and unstable picture were the symptoms. Check the vertical height and linearity controls for misadjustment. This same condition may also be found if the HV at the CRT anode is too high. Light and dark bands appear across the screen, running from top to bottom, indicating a portion of the horizontal sweep (15 kHz) signal is finding its way into the video amplifiers. The probable cause is poor lead dress, most frequently caused by a horizontal sweep pulse feeding into the delay line. Change the lead dress.

PHILCO 15M90 CHASSIS
Narrow picture

90

The width on these sets and all other sets using the 21-inch round picture tube may be slightly narrow, otherwise the picture is normal. Check the horizontal frequency and sinewave coil.

In some of these chassis the horizontal hold control will cause the hue to shift or change. Check for defective duodiodes in the horizontal sync comparer.

If oscillation appears in the color picture or color streaks run across the screen horizontally. Check for a poor connection in the bandpass amplifier, usually in the plate transformer.

When HV arcing shows up as an interference pattern on the TV screen, suspect the 1K resistor in series with the HV lead or a defect in the plate lead and socket on the 6BK4. Replace only with Philco part number 41-4367-1, socket and lead assembly.

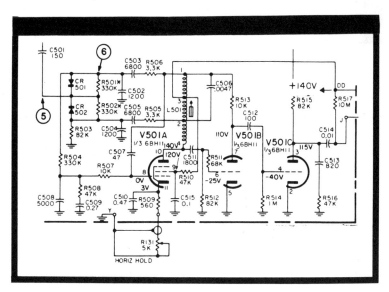

91
PHILCO 15M90 CHASSIS
Picture pulled in from top and bottom

This particular color set had insufficient vertical height and also displayed some picture foldover.

Should the trouble be insufficient height with complete foldover, the trouble is usually an open 80-mfd filter in the plate supply of the vertical output tube. If there is a loss of sweep, the trouble may be an open 1000-ohm resistor in the vertical output tube (V14) grid circuit. If the height is insufficient, the fault is usually either a defective booster-boost diode D7 or an open or increased value R164 (2200 ohm, 3w), which completes the vertical output cathode circuit to ground. These circuits are shown in Fig. 72.

92
PHILCO 15M91 CHASSIS
Color bars through picture

This is usually an intermittent condition, with wide color bars (generally red) drifting through the picture vertically. Other symptoms are poor sync, loss of color, picture may go black, tube filaments go out and the picture is of poor quality.

All of these symptoms may be caused by poor grounding between chassis projections and grounding eyelets on the PW panel. The symptoms in each case obviously depend upon the panel involved. To correct, solder a wire between the grounding point on the panel bottom and bring it out to a good chassis ground point. Resoldering of the eyelet ground may not always be a very satisfactory solution.

FIGURE 72

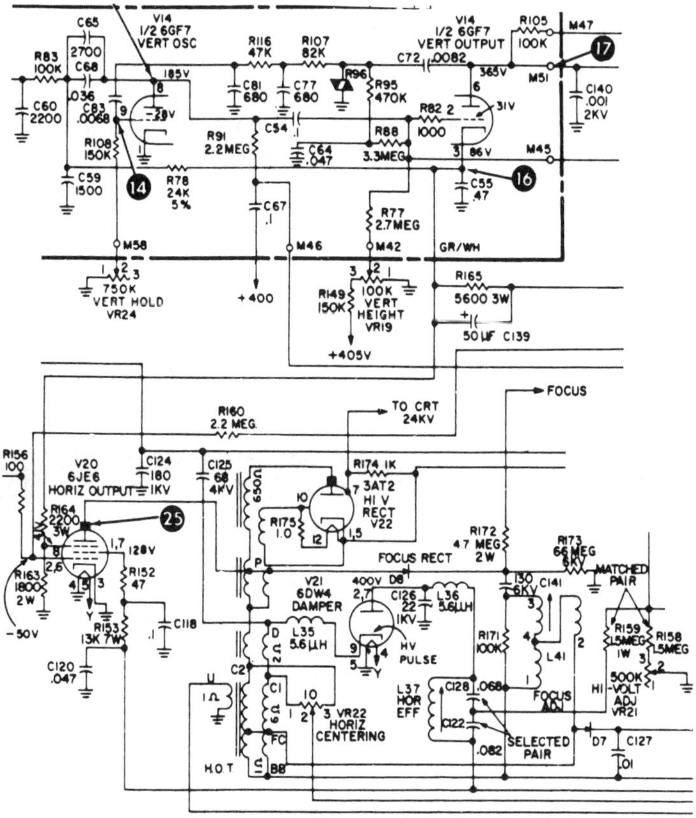

111

93   PHILCO 17MT80 CHASSIS
Vertical jitter or poor interlace

This color chassis had developed two different vertical problems. One was a vertical jitter and loss of sync. The other trouble was a poor vertical interlace symptom.

First let's look at the vertical jitter or loss of vertical sync. In some cases the vertical deflection was completely lost. The trouble was a faulty (.0033 mfd) capacitor, C123, located in the feedback circuit from the plate of the vertical output tube to the grid of the vertical oscillator. Refer to the schematic in Fig. 73.

The other symptom was poor vertical interlace. The picture also had black-and-white vertical bars (horizontal interference) and no video on strong stations. The probable cause is a defective capacitor C1 (3.9 pf) found in the base circuit of transistor (Q3) the AGC amplifier stage.

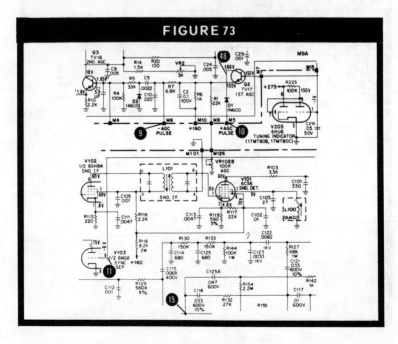

FIGURE 73

PHILCO 18MT10-B CHASSIS
Hash or weak negative picture

94

This chassis had hash and dark streaks, mostly horizontally, across the screen, which may be due to the misadjustment of AGC control (VR2) on the IF panel, Fig. 74. The IF AGC with no signal (AGC gate inoperative) will be about 3.2 volts after proper VR2 adjustment. Note: Place the channel selector knob between channels to obtain a no signal condition.

This color receiver produced a very weak negative picture with oscillation hash. The trouble is caused by a reverse bias on the third IF stage. Certain external AGC voltages applied to this stage may give a positive picture of considerably less than the 1.5 volts normal peak-to-peak signal that should be found at the video detector. The probable cause for this picture symptom is an open 3-pf, neutralizing capacitor in the second IF stage. The Q5 transistor stage shown in the schematic of Fig. 74.

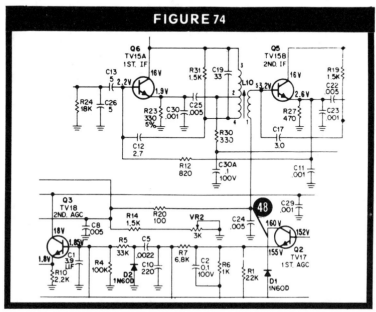

FIGURE 74

95

PHILCO 16QT85 CHASSIS
Picture hash

This set had some sound interference in the video on color programs (hash appeared on the screen when the fine tuning was adjusted toward the sound side for best picture).

For this problem make the following checks: Refer to the schematic in Fig. 75. The 41.25-MHz trap (L6 top) may be misadjusted.

The AGC voltage may also be incorrect. Check for a defective third IF transformer or faulty video detector diode.

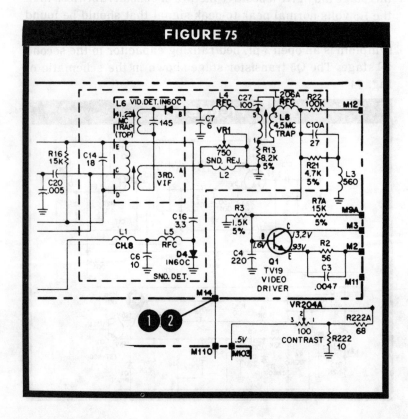

FIGURE 75

PHILCO 16QT85-A CHASSIS
Dark bars or bands on raster

96

This set produced a single dark hum bar that was either stationary or drifted vertically across the picture.

To cure this problem replace capacitor C209C, 200 mfd, 25v, across the brightness control in the cathode of the video output tube (12GN7) shown in Fig. 76.

Some of these same sets may have vertical bands or lines on the left side of the raster.

To correct this, properly dress the orange B+ lead coming from terminal strip B6 (lug 5), going past the IF panel along the front chassis apron, and terminating on strip B4 (lug 5). This lead should be dressed away from the IF panel toward the front apron and down close to the bottom of the chassis. Also check filter capacitor C208A.

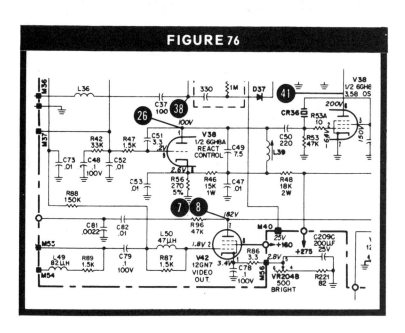

FIGURE 76

97 PHILCO 17QT82 CHASSIS
No raster or high voltage

Check the horizontal output transformer and associated circuitry. A defective focus coil or damper tube may cause the sweep transformer to fail. Check the anode cap dust cover. Carbonization of the anode cap can lead to CRT and sweep system failures. Check and clean the HV button area.

With no raster (raster is present when the service switch is in the service position), check the plate voltage of the video output tube and L51 (video peaking coil). Note schematic in Fig. 77. The open plate circuit will cause the CRT cathode voltage to increase to the point of CRT cutoff. Also check the video output (12GN7), it can cause the same symptom.

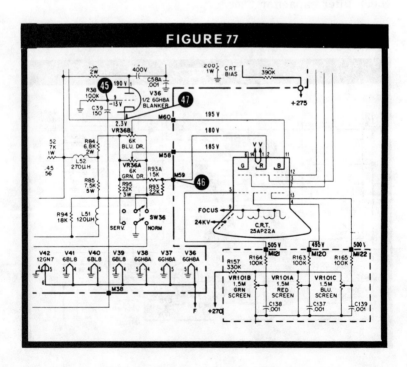

FIGURE 77

PHILCO 18QT85-A CHASSIS
No color or random color

98

If there is a good black-and-white picture, but no color, replace the 3.58 MHz oscillator tube (V38), check the crystal (CR36), plate voltage, transformer L42, and associated oscillator circuits. See the schematic for this color circuit in Fig. 78. Note: When checking for color, always make sure you have good B&W reception first.

If there is random color reception, this indicates the 3.58 MHz oscillator is operating but not on frequency. After tubes have been checked or replaced and the trouble still persists, check resistance coil L39. To adjust L39, short TP39 to ground and adjust the reactance coil for color sync zero beat. Replace the crystal if drift is beyond the control range. With your scope, check the blanker pulse, which can affect the bandpass amplifier and the phase detector stages.

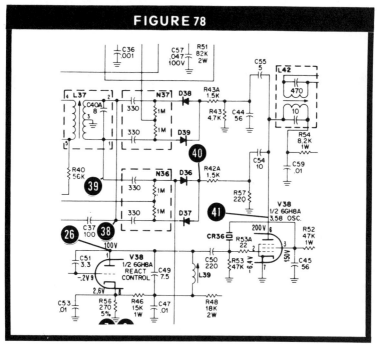

FIGURE 78

PHILCO 19QT85-B CHASSIS
Loss of color or improper sync action

Refer to Fig. 79, the color sync circuits for these following problems:

If the color picture is slow to come on and the color sync drifts, it may be due to improperly matched or faulty diodes D38 and D39 in the color killer circuit. The DC voltage at the N37 phase detector output must be within 0.2v of zero when tuned to a snowy raster.

A loss of sync, improper color phase and where color is slow to come on could be caused by improperly matched or defective diodes D36 and D37 in the color phase detector. The DC voltage at the N36 phase detector output must be within 0.2v when tuned to a snowy raster.

Where color sync is missing and color is also slow to come on, check coil L39 (open). Replace the coil and check the color alignment.

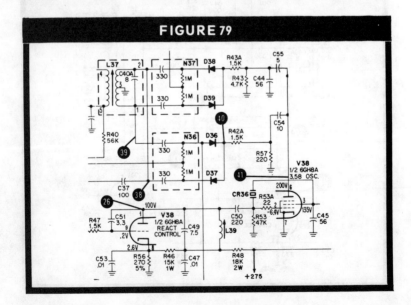

FIGURE 79

PHILCO 19QT87-B CHASSIS
"Odd Ball" mixed up color screen

This color receiver had a lot of mixed up looking colors on the screen. There was a red ball in the center with streaks of color extending from it which looked like a flying comet. The bottom of the screen also looked a little hazy. The little culprit that caused all of these odd colors was a 0.1 mfd blanker-coupling capacitor shown as C43 in the schematic Fig. 80. This coupling capacitor feeds the blanking pulse into the cathodes of all three color-difference amplifier tubes. The capacitor (C43) was found to have a slight leakage. Should this capacitor short, it would bias-off the three color-difference amplifiers and no raster would be seen. This same problem may also be found in other brands of color sets. This fault will certainly cause some odd picture symptoms.

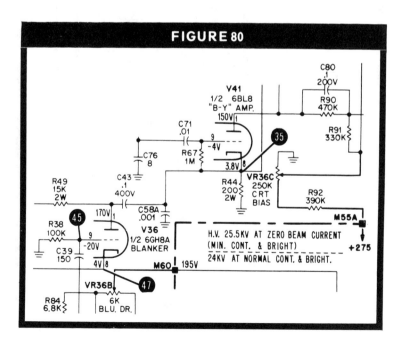

101 PHILCO 19QT87-B CHASSIS
A shifty picture

The color set had a good picture, but the raster was too far to the right of the screen by about 3 inches. It would continue to shift to the right as the brightness control was increased. In some cases the picture was too far to the left of the screen.

This picture shift condition was caused by a faulty horizontal centering diode, D206. This is shown in the horizontal sweep circuit of Fig. 81. If the picture is too far to the left, short out diode D206. However, do not remove D206 from the circuit. In some severe cases, install a 5 to 8 ohm resistor in parallel with diode D206, as this will help move the picture back to the right.

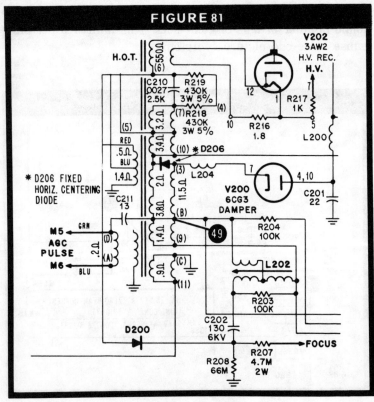

PHILCO 20AT88-B CHASSIS
Poor horizontal sync or horizontal hold

102

For a chassis that has poor horizontal hold, make the following checks: Look at the leads going into the body of capacitors C129 and C130 to see if they have been damaged by heat from a soldering iron. Both capacitors may check good, but if there are signs of damage to the leads from too much heat, they may be opening up intermittently, which will show up as an intermittent horizontal hold. Refer to Fig. 82 for the horizontal oscillator circuit.

For poor horizontal sync, check capacitors C142 and C143 in the phase comparer circuit to make sure of the values (120 and 82pf, respectively). We came across a couple of these chassis where they had been interchanged during production. Also check the phase comparer diode D100 as it may be defective.

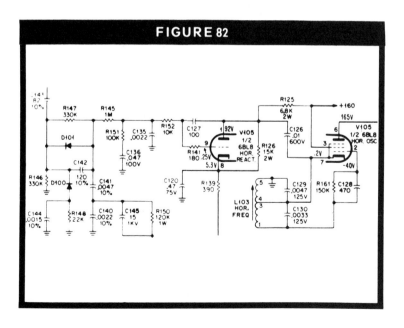

FIGURE 82

103
PHILCO 20QT90-B CHASSIS
No color

For chassis that have no color or may receive color intermittently, check all transformers, L37, L40, and L41 for open or shorted windings. Refer to the schematic in Fig. 83. Note: For no-color problems, be sure to override the color killer operation to determine if the trouble is in the chroma, color sync, or 3.58-MHz oscillator circuits. If there is no chroma, the trouble is in the bandpass or second chroma amplifier. If the trouble is a loss of sync, the fault is in the color phase detector, burst amplifier, reactance stage, or 3.58-MHz oscillator. If the color is primarily magenta with some green, the 3.58-MHz reference voltage is not being fed to the demodulators. Use test point lug 49 for isolating trouble between the bandpass and second chroma, as well as isolating color sync trouble ahead or behind the color phase detector.

FIGURE 83

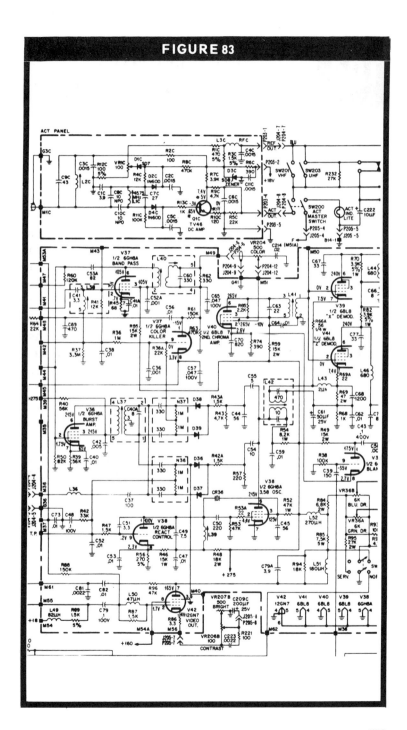

104 RCA CLC15 CHASSIS
Intermittent raster fade-out

This color set had an intermittent picture fade-out problem that would show-up after the receiver played about one or two hours. Of course, during the three service calls that we made, the set performed perfectly. The chassis was pulled and connected to the test jig and allowed to cook. After many hours of "air-tests" and circuit checks, the raster did fade out, but it was noted that the HV was correct. The prime suspect was now the (chroma) PW700 circuit board. The trouble was found to be a cold solder ground eyelet connection on one corner of PW700. This is actually the ground connection for R742 (see schematic in Fig. 84) a 390K resistor in the control grid circuit of V707B, the blanker tube. This caused an upset of bias for the blanker tube and placed the CRT into cutoff, thus a blank screen. Other RCA chassis have developed these same cold solder ground problems. A good service tip, should you run across a chassis with odd-ball or intermittent trouble symptoms, would be to resolder all of the circuit board ground eyelet connections. This may well solve a lot of those troublesome "bow-wows."

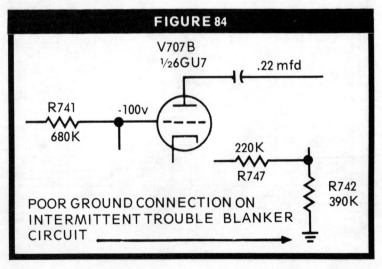

FIGURE 84

RCA CTC16X CHASSIS
Distorted or compressed scanning lines

This RCA set had a very unusual picture scanning line problem. The raster had a small amount of hum in the vertical sweep which caused a slowly moving pattern of compressed scanning lines at one part of the screen and expanded lines at another. Refer to the vertical deflection output circuit in Fig. 85 for the solution to this "odd-ball" symptom. It was also noted that R162 was operating very hot. This problem can be caused by faulty (open) C136A or C118A filter capacitors. Filter capacitor C118A, located in the power supply stage, is the main capacitor for determining the B+ voltage. When it is open, C136A is inadequate to fulfill the function of the input capacitor, so the insufficient filtering allows ripple in the B+ to reach the vertical sweep output tube, and the capacitor charging current thus overheats resistor R162. Check or replace the filter capacitors and R162.

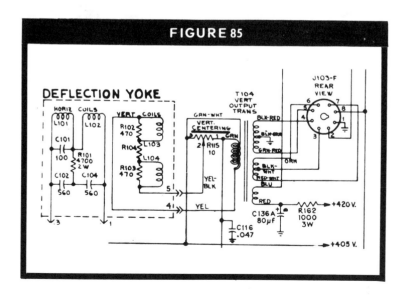

106

RCA CTC16X CHASSIS
Blurred picture and low high voltage

The raster on the screen of this RCA color receiver was dim and out of focus. The high voltage at the CRT anode cap measured 18 KV. All tubes in the horizontal sweep section checked out good and the HV regulator was operating properly. After the set operated about 10 minutes it was noted that the plate of the 6JE6 sweep output was turning red. The horizontal drive voltage at the control grid of V105 measured -42 volts, which is too low. It should be -56 volts as shown in Fig. 86. We completely checked the horizontal oscillator stage (V502). It was found that R544, a 1.5 megohm resistor from pin 7 of the horizontal oscillator to one side of the horizontal hold control, had opened. This caused the low horizontal drive to the grid of V105 and hence the red plates.

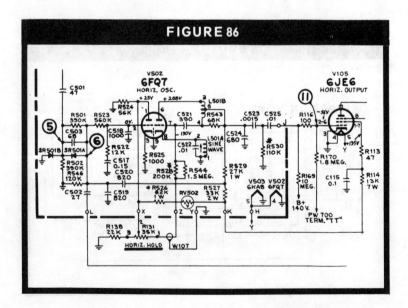

FIGURE 86

RCA CTC17X CHASSIS
Color confetti on B & W picture
107

This set produced color confetti that was evident on the black-and-white picture regardless of the color killer control setting. The color killer circuit for this chassis is diagrammed in Fig. 87. A good way to begin a logical analysis is to measure the DC voltages in the color killer, phase detector and killer detector circuits. Then use your oscilloscope to check for correct or incorrect waveform pulses on the color killer detector, V705B, and color killer stage, V701B. The fault for this symptom was caused by a shorted C702, .047 bypass capacitor unit. This is in the plate circuit of V701B. Also be on the lookout for a defective (gassy) color killer tube, V701B. If this tube becomes slightly gassy it will cause a positive voltage on its own control grid.

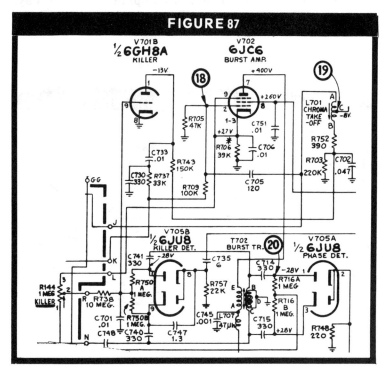

FIGURE 87

108

RCA CTC21 CHASSIS
Dark screen, no raster

There was no brightness on the CRT of this receiver but the sound was normal. The HV was low, about 12 KV at the CRT anode. When the socket of the CRT was pulled, the HV went back up to around 26 KV, indicating CRT or CRT bias problems. In this chassis the CRT bias was improper due to an open transformer (T106) in the video circuit shown in Fig. 88. When the HV builds up and then drops off, it could be CRT trouble, a soft HV rectifier or faults in the CRT bias circuits.

If the CRT grids are not positive enough, look for trouble in the chroma section. And if the CRT cathodes are more positive, look for a fault in the video amplifier stages. Always check the CRT bias voltages before condemning the picture tube.

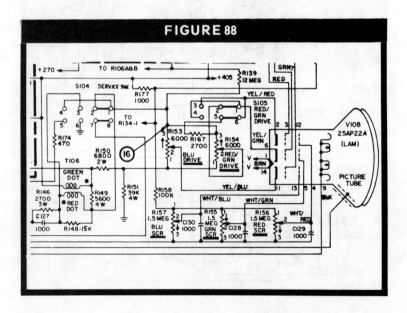

FIGURE 88

RCA CTC21 CHASSIS
Blurry picture

This RCA color chassis came into the shop with a very blurry picture and low screen brightness. No action on the raster was noted when the focus control was adjusted. The high-voltage at the anode of the CRT was low and the HV adjustment control also produced no action. All of the tubes checked good.

Many component faults could cause these same symptoms, but this chassis had a shorted 6KV capacitor in the focus voltage control circuit. This is a 130 pf, 6 KV capacitor shown as C107 in the schematic, Fig. 89. Replace C107, set the high-voltage to its correct value and adjust the focus transformer (T101) for a sharp picture.

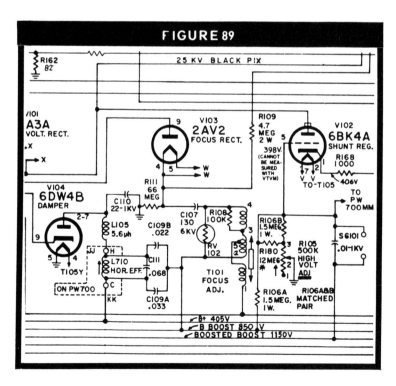

110 RCA CTC21 CHASSIS AND MANY OTHER RCA MODELS
Focus problems

Use the focus circuit in Fig. 90 as we go over these service hints.

1. Focus transformer T101, if defective, will cause a poor focus condition. Check to see if the windings are burnt or shorted.

2. The 130-pf capacitor (C111) is the input filter. If this opens, the focus voltage will be low. If C111 shorts, the focus voltage will measure the same as the B-boost.

3. The 66 megohm resistor (R116) provides a load when the CRT is not drawing current. The picture focus will be erratic without it or if it goes up in value.

4. The 4.7 meg resistor (R117) protects the other components from arcs. Older RCA chassis use a 1 meg, but this should be changed to a 4.7 meg for better protection.

5. RV102, a varistor, is another protection device, particularly against damage that could be caused by a shorted focus rectifier tube.

6. The 100K resistor (R121) reduces transformer (T101) ringing. If it is burnt, the focus tube (V101) should be changed.

The total focus voltage change is about 600 volts. Thus, the focus voltage must be around 20 percent of the HV for best performance. Remember, either the focus voltage or the HV can cause a poor focus symptom. Check both to be sure which is correct. A weak picture tube also will not focus very sharp, but if a "sharper" focus is noted on either side of the focus control adjustment, the focus voltage must be performing correctly.

FIGURE 90

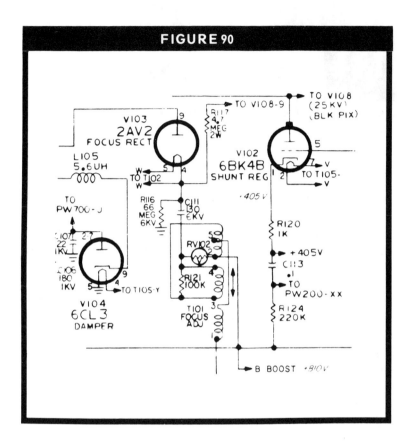

111
RCA CTC21 CHASSIS
Hum or buzz in the audio

To locate the trouble, unplug the speakers and turn the volume down. Any buzz still heard may be from a transformer, the CRT purity shield or metal bottom plate.

Hold the purity shield. If the hum stops, pack sponge rubber between it and the CRT. Hold the bottom plate. If this reduces hum, pack sponge rubber between the filter choke and the bottom plate.

Move the set-up normal switch to the "set up" position. If the buzz is gone, the vertical output transformer is faulty. If the hum is still heard, it is the power transformer.

Next step is to connect the speakers. If the buzz is heard, the problem is in the sound circuits that are shown in Fig. 91.

Now move the set-up-normal switch to "raster." If the buzz is gone, there may be an overload in the picture IFs or video stages, or a defect in the sound detector or sound IF amplifiers.

If the buzz is still there, but is not changed by volume control adjustments, a B+ filter may be defective.

However, if the buzz is still heard, but is worse near the center of the volume control range, the buzz is being picked up capacitively by the volume or tone control wiring. Check the cable and ground connections.

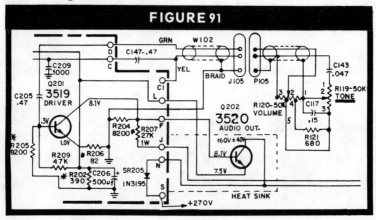

FIGURE 91

RCA CTC22 CHASSIS
Verltical and horizontal sync instability 112

This small screen RCA color receiver displayed intermittent horizontal and vertical sync. The vertical and horizontal oscillators were operating at the correct frequencies and the picture looked as if there is an AGC trouble. A few quick voltage checks were made around the sync and AGC circuits and the plate voltage was found too low. In the search for this low voltage, a defective resistor was found. It was R412, 680 ohms, which comes off the 280-volt B+ supply and feeds voltage to the sync and AGC tube stages. Also, in some cases of this trouble, the brightness control acts like an AGC control. Replace R412 with a 680-ohm 2-watt resistor. See the schematic in Fig. 92.

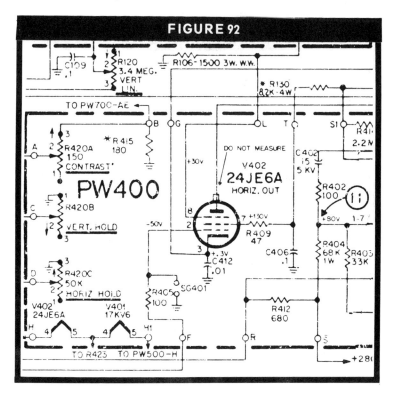

113 — RCA CTC22 CHASSIS
Scalloped picture

This portable RCA color set had good sound but produced a scalloped picture that had a Christmas tree effect. The schematic for this horizontal control circuit is shown in Fig. 93. The dual silicon AFC diodes (CR501) checked out in good condition. However, this type of picture trouble usually occurs in the anti-hunt network of the horizontal control tube, V503B. The network in this chassis consists of C514, C513 and R254. The best way to check this circuit is to sub-in known good components. This was quickly done and C514, a .15 mfd capacitor, was found to be defective.

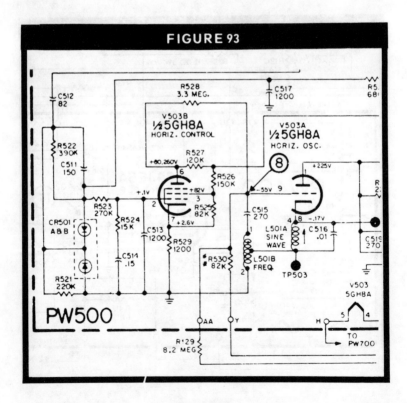

FIGURE 93

RCA CTC22 CHASSIS
No vertical sweep

This small screen color set had good sound, but the vertical sweep was lacking. The vertical sweep tube, a 10GF7A, was good so we dug into the vertical sweep section. In short order the VOM found the plate voltage missing from pin 8 of V502B, the vertical oscillator section. Refer to Fig. 94 for the schematic of this circuit. The loss of plate voltage was caused by a shorted bypass capacitor, C109, that comes off of the center arm of the vertical linearity control, R120. Replace C109, a .1 mfd capacitor, and check to see if the vertical linearity control was damaged. It may have a burned spot on it. Then, check and adjust the picture linearity if needed.

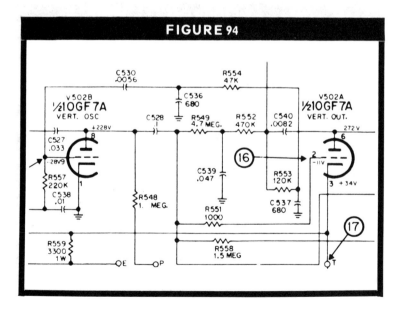

FIGURE 94

115
RCA CTC24 CHASSIS
No color

This receiver was producing a clear black-and-white picture and good sound, but there was no color. So some routine checks were made in the chroma section and the 3.58 MHz oscillator circuit was a prime suspect. However, the color subcarrier oscillator was working okay. A few more tests revealed that the color killer circuit (note schematic in Fig. 95) was at fault and was turning off the chroma amplifier stage. The trouble was a leaky C726, .01 mfd capacitor at pin 9 of the color killer stage, V703A. Thus, a new C726 capacitor cured the no color symptom for this set.

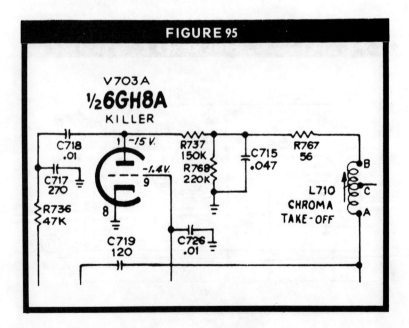

FIGURE 95

RCA CTC24 CHASSIS
Raster will not focus

116

This problem has been found on many makes and models of color TV receivers. If the HV at the CRT anode checks at 25 KV and the focus voltage will vary from about 4500 to 5500 volts, try this: Pull the CRT socket off and look at the hole of pin 9. If you see a greenish corroded powder, it is the cause of your focus problem. Pin 9 of the CRT will probably look the same. And if you pull on the black lead it may come out of the socket. These focus pins will sometimes corrode like this. Thus, you end up with a very poor contact, and in some cases the socket will be eaten away at this point. Note the socket drawing in Fig. 96. Change the CRT socket assembly, check the HV and adjust the focus control for sharpest raster lines.

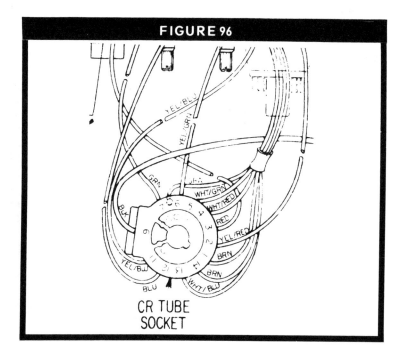

117

RCA CTC25A CHASSIS
Sound troubles

A defective transistor caused the sound to "motorboat" in this RCA color set. (See the schematic for these audio stages in Fig. 97.) Excessive base-to-emitter leakage in Q202 caused this motorboating condition. It should be noted that the drive stage has a gain of about 2, while the output transistor has a gain of around 2,000. Transistor Q201 functions almost as an impedance matching device.

Another one of these sets came into the shop with no sound or picture. In this case a good tip was the smoke coming from the sound PC board. The trouble was a shorted Q202 audio output transistor. Also check diode SR205 and R202, a 390-ohm resistor.

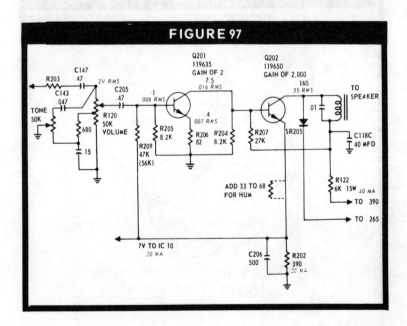

RCA CTC25X CHASSIS
No video or weak and washed out picture

118

This particular RCA color chassis had developed a very weak, washed out picture. The picture tube displayed a lot of brightness and the sound was good. The video amplifier stages were the prime suspect and this assumption proved to be correct. A few voltage measurements were taken in the video circuits (shown in Fig. 98), but all were within tolerance. The oscilloscope was then put to use to trace the video signal. The fault was located in the first video stage, V304A. The 22-ohm cathode resistor, R326, had opened up. Also, replace V304A, type 6LF8 tube as it may have shorted and caused the cathode resistor to open up.

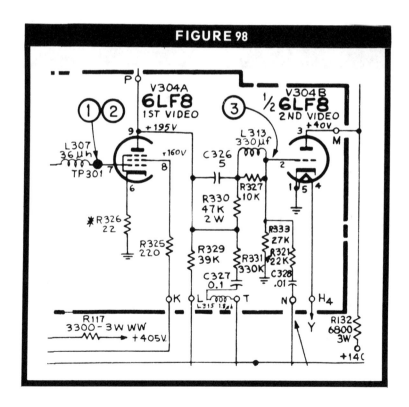

FIGURE 98

119 RCA CTC25X CHASSIS
No raster

The screen was dark on this receiver and there was no high voltage being developed. The drive signal at the control of the horizontal output tube was correct. All of the routine checks in the damper stage, yoke and boost circuits proved fruitless. The culprit this time was a shorted pincushion phase transformer, T107, which is shown in Fig. 99. The only sure way to solve this problem is to replace T107. Should you disconnect T107 for a check, this chassis will still not produce any HV or horizontal sweep. In some troubleshooting circumstances, substitution is the only way to go.

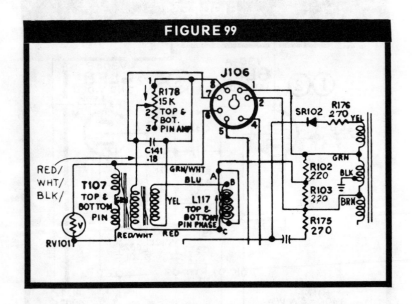

RCA CTC25X CHASSIS
No HV or raster

120

This HV and horizontal sweep malfunction was a tough one to isolate. The raster did not appear on the screen and it was noted that the sweep transformer was warm to the touch. Of course, the B boost voltage was low. The drive signal waveform and screen voltage were normal for the horizontal output tube, V105. (Refer to the horizontal sweep output circuit in Fig. 100.). All of the routine checks in the damper, yoke and boost stages did not uncover the fault. After a few more checks, the problem was diagnosed as a shorted AGC pulse coupling capacitor, C113, that connects to pin 4 of the sweep transformer T102. This was loading down T102, thus eliminating the horizontal sweep and high voltage. The loading caused the transformer to overheat, too. I can recall another chassis with this same trouble, where the AGC lead wire was shorted to the chassis, thus causing this same symptom.

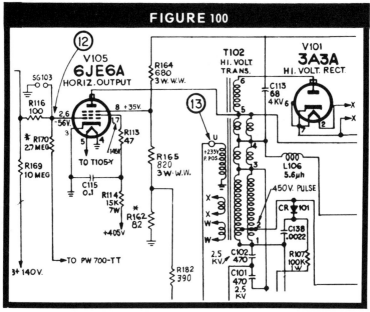

121 RCA CTC25X CHASSIS
Poor picture lock-in

This receiver had a good clear picture but had a very intermittent and unstable sync problem. With this picture information we began our circuit investigation in the sync separator and noise inverter section. A few voltage checks were made but all appeared near normal. Since the oscilloscope is the most useful instrument to troubleshoot the sync stages, we put it into action around these circuits and found the problem in short order. This sync trouble was caused by a leaky C527, a .0033-mfd capacitor which was putting a positive voltage on pin 2, control grid of the 6KA8 sync separator. Refer to the sync separator circuit in Fig. 101. Replace C527 and adjust the AGC control level.

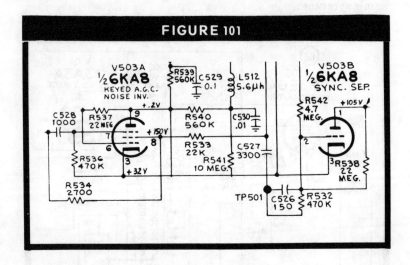
FIGURE 101

RCA CTC27X CHASSIS
Gray scale shift

This TV receiver was operating just about normal, except that the color intensity control had limited range after warmup. The gray scale of the CRT would shift toward bluish-purple and also varied with the color intensity control level. A color bar pattern was fed into the set and a scope was used to signal trace through the chroma stages. We could not pin point the fault, so an alignment was attempted for the chroma bandpass stages as shown in Fig. 102. It was noted that T701 would not tune properly. The transformer was replaced and the chroma stages aligned correctly. We now had a very good color picture with no color shift. A fast way to track down some troubles is to make an alignment check.

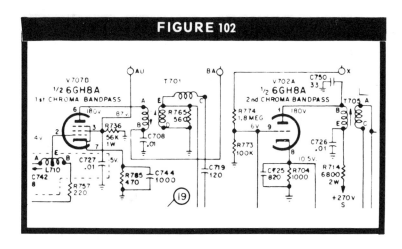

FIGURE 102

123

RCA CTC31 CHASSIS
Color fades out

This RCA receiver would operate fine for about 20 minutes, then the color would fade out. And this was a complete loss of color. The fine tuning and color killer adjustments were checked and found to be correct. The 3.58 MHz color oscillator was operating and a few voltage tests in the color section did not present a clue. This fault appeared to be in the chroma bandpass amplifiers and the oscilloscope was then used to signal trace through these stages. The trouble was found in the 2nd chroma bandpass amplifier transformer, T705 (see Fig. 103). A cold solder joint was found at terminal A of this transformer. All of the terminals for T705 were resoldered and normal color reception was restored.

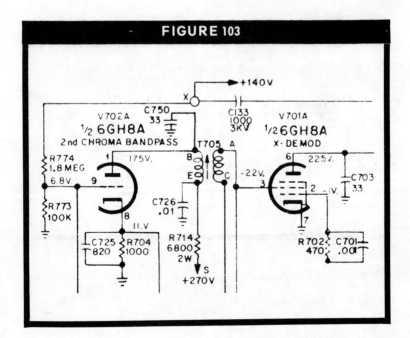

RCA CTC31 CHASSIS
No color

A very good possible cause for no color from this chassis is a fault in the color killer stage shown in Fig. 104. The color killer transistor, Q701, operates as an open or closed switch. Its base voltage comes from two sources, the variable negative voltage from the 3.58 MHz oscillator grid and a positive voltage from the killer control. This no color condition was caused by a shorted Q701, 3533 color killer transistor.

Do not ground the collector to see if the color returns, for this will result in a zero demodulator screen voltage. Some very poor quality color will be the result. A better way is to attach a clip lead from emitter to collector (with set turned off) to see if the color returns. If the color returns, the color killer control is misadjusted or a defect exists in the killer circuit.

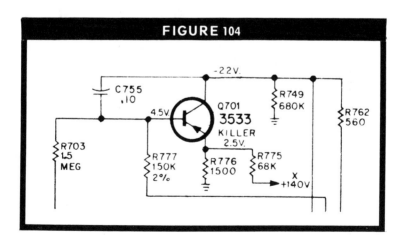

FIGURE 104

125

RCA CTC31 CHASSIS
Washed out picture

This large screen color chassis had a washed out, but overly bright picture and the color hues were not correct. However, the gray scale of the raster appeared to be tracking properly. The picture tube checked good and all color set-up adjustments were in good order. The set owner said that the set snapped and popped very loud just before all of this trouble developed. This picture symptom developed when all three clamp diodes in the color amplifier section blew out. (See the schematic in Fig. 105.) This was caused by an arcover in the 6BK4 HV regulator tube. Replace the three faulty diodes CR702, CR703, and CR704. Replace the 6BK4 regulator and adjust the HV to 25 KV. You should now have a good color picture with lots of contrast.

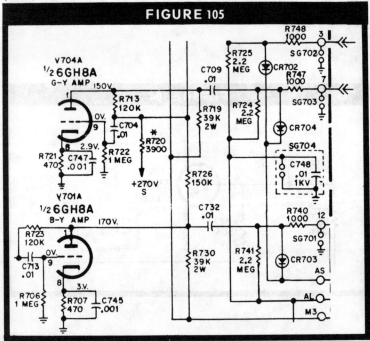

FIGURE 105

RCA CTC31A CHASSIS
Low brightness

126

The raster on a CTC31A chassis was very dim. You could just make out the program image on the screen, and the brightness control had no effect on the picture.

Because the HV reading at the CRT anode was correct, the trouble was probably in the video section. Voltages on the video amplifier were higher than normal (see Fig. 106) and a new 12HG7 tube didn't help.

An ohmmeter check revealed that contrast control R126A was open. Probably the video amplifier had shorted and burned a spot on the 100-ohm control. If the contrast control was turned to the extreme end of its rotation, a bright picture could be seen. Make sure you replace both the tube and contrast control.

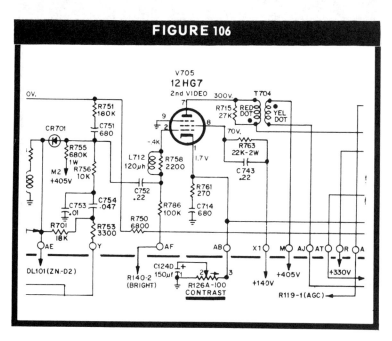

FIGURE 106

127

RCA CTC38 CHASSIS
Bright, washed out picture

The picture was faint and there was no control of the brightness, but the sound was good on this RCA color receiver. At first glance the set appeared to have a defective CRT; however, it checked out good for this lucky set owner. All of the bias voltages for the CRT were found in good order. Next, the three stages of video amplification were tested, and no fault was discovered, but the scope found transformer L710 in the third video stage, V707 (see Fig. 107), to be faulty. The transformer secondary winding had opened. You may be able to repair L710 in such cases; if not, a replacement is called for.

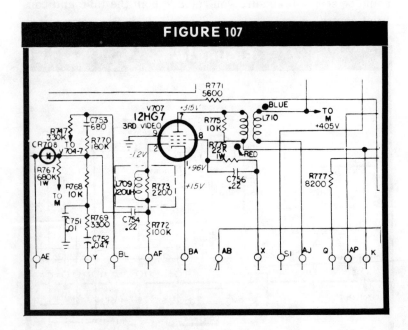

FIGURE 107

RCA CTC38 CHASSIS
Weak color and loss of color sync

128

Weak color symptoms associated with the color sync and-or AGC-killer circuitry in these chassis may be the result of an open winding in the first bandpass transformer, T701 (RCA part number 124761). Refer to Fig. 108 for the 1st color bandpass circuit schematic. If the open occurs in the winding connected between terminals C and E, the color signal input to the burst amplifier stage will be incorrect. However, the path for the chroma input to the second bandpass amplifier stage will be normal. Use a scope and color-bar generator for troubleshooting this type of trouble.

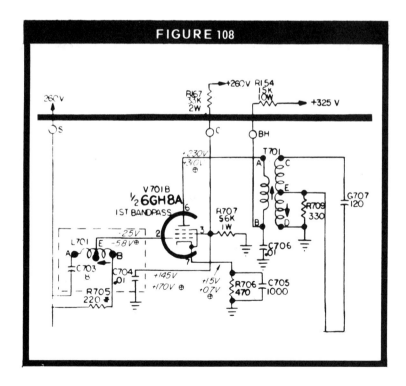

FIGURE 108

129 RCA CTC40 CHASSIS
Excessive brightness

This solid-state chassis had a picture that was very bright and could not be turned down with the brightness control. A good way to start checking for this type of problem is to measure all of the bias voltages on the CRT. This brightness trouble was located in the horizontal blanking stage (Q719) and kine bias circuits. Refer to the schematic in Fig. 109. The problem component was a faulty zener diode CR712 which is located across the CRT bias control. This zener diode may develop leakage or be shorted. Replace CR712, set the CRT bias and check the high voltage at the CRT anode.

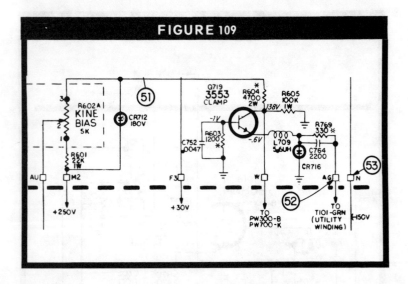

RCA CTC40 CHASSIS
Video "smear" and loss of sync

130

A wide variety of video and-or sync symptoms in this chassis may be the result of an electrolytic capacitor (a 20 mfd, 15v electrolytic, RCA stock number 121995) changing value.

Possible symptoms include: Video smear, video bends or wave, unstable sync, or various combinations of these symptoms. In addition, the symptom may vary with the brightness control setting. This capacitor is shown as C354 in the schematic of Fig. 110 and is in the base circuit of transistor Q305.

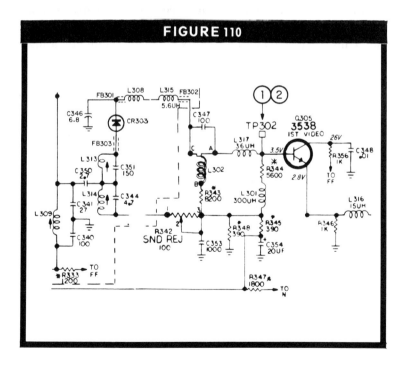

131
RCA CTC40, 47 CHASSIS
Piecrust or geartooth effect

The .068 mfd commutator capacitor used in the horizontal deflection circuit of this chassis is located either on the PW300 board or on a terminal board under the SCR heat sink. Refer to Fig. 111. Two capacitors in parallel are never used. If the capacitor is on PW400, it is C403; off the board, it is C123. In either case, the capacitor is the same type and is RCA stock number 165437.

A "piecrust" or "geartooth" effect in the raster (scalloped edges) at different brightness levels may be the result of an open R423. Under these conditions the high voltage and brightness limiter operation appear to be normal, while in some instances the symptom may be accompanied by a high-pitched squeal.

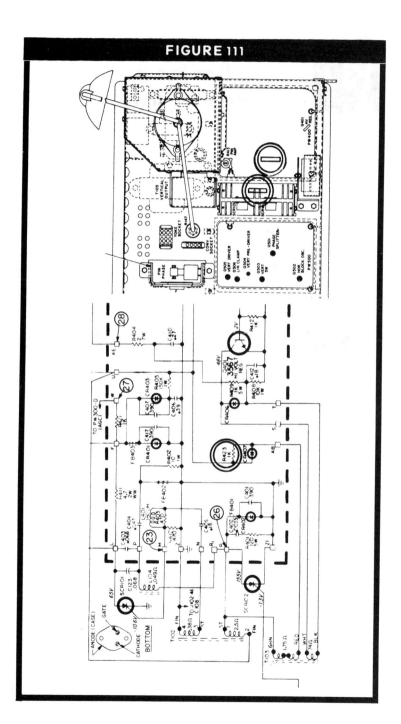

FIGURE 111

132
RCA CTC44-CTC47 CHASSIS
Hum bars and-or degaussing interference

A defective resistor R104 (Note Fig. 112) in the 240v DC supply circuit may cause this problem. Should this resistor fail, the color receiver may continue to operate on the 220v DC supplied by the degaussing bridge rectifier output. Symptoms for R104 failure may include a hum bar across the screen or degaussing "interference" in the raster.

For a quick check, unplug the degaussing coils and if the video is lost, the 250v DC supply is inoperative. Replace with a 47 ohm, 4w flame retardative type resistor (stock No. 132879).

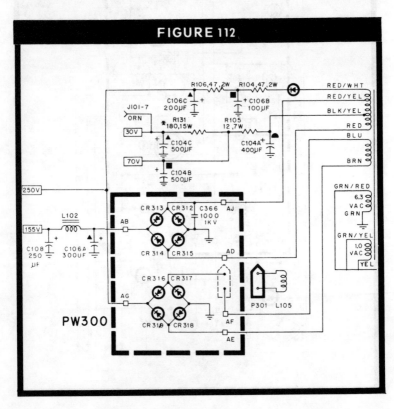

FIGURE 112

RCA CTC46-49 CHASSIS
High voltage troubles

133

The HV may be too high or too low and cannot be adjusted. The following may solve this problem for you.

In this chassis the side-pincushion circuit operates in conjunction with the HV regulator, and a failure of the pincushion amplifier will normally affect high voltage. If the HV is excessive and uncontrollable, the cause is either a shorted pincushion amplifier or an open regulator transistor. If the HV is low and uncontrollable, look for an open pincushion amplifier or a shorted regulator transistor.

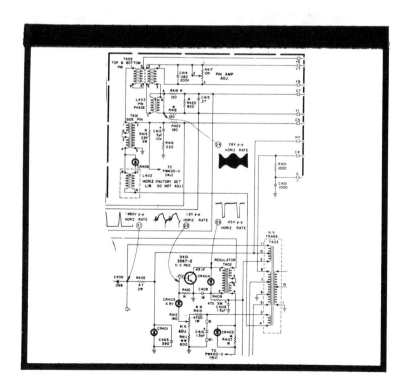

134 RCA CTC52 CHASSIS
Picture "hum-bar" interference

Some of these sets may show a hum-bar in the picture, when operated in weak signal areas. The interference appears as the "silicon-rectifier-type-bar" usually associated with the power supply diodes. In these chassis the interference may be caused by the neon channel indicator bulbs. Absence of the bars when the supply voltage is disconnected from them indicates the bulb as the source. One way to eliminate this interference is to add a 680 pf ceramic to ground as shown in the drawing of Fig. 113.

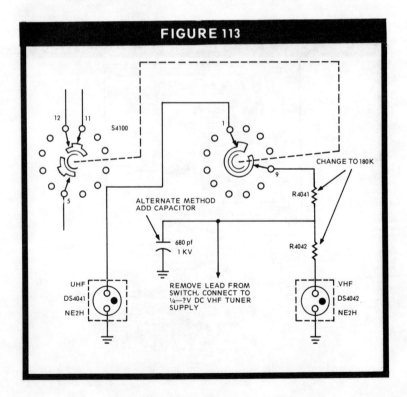

ZENITH 14A9C29Z CHASSIS
Color fades out

This TV receiver had a perfect color picture and sound but the color would go out after operating about 30 minutes. We jumped a clip-lead across test points K and KK and lots of colored stripes then appeared across the screen. This indicated the color amplifiers and demodulators were working, but the 3.58 MHz oscillator was off-frequency. Next, test point W was shorted to ground in order to zero-beat the 3.58 MHz oscillator. However, the oscillator was now on-frequency which pointed to probable trouble in the AFC phase detector or color burst amplifier stage. We applied some circuit cooler on several components in the phase detector circuit and capacitor C327, a dual 2X .001 mfd unit, was found to be faulty. This capacitor would open after it warmed up, which then unbalanced the AFC phase detector and changed the control voltage on the grid of the reactance control tube, V214A. See Fig. 114 for this AFC phase detector circuit schematic.

This is a "rare" and very tough to find intermittent trouble. In some of the other Zenith chassis this same component has produced an intermittent color fade-out symptom. Replace capacitor C327 and check for proper color alignment.

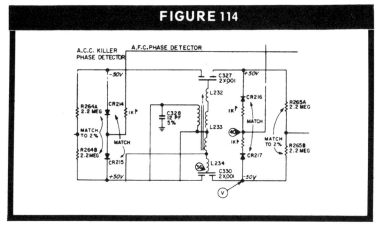

FIGURE 114

136
ZENITH 14A9C50 CHASSIS
Diagonal line from corner of the screen

This color chassis had a dark, waving diagonal line coming from one corner of the screen. We called this trouble the "dancing string caper." This "dancing string" was caused by a defective transistor, Q204, in the pincushion correction circuit shown in the schematic of Fig. 115. The "tip-off" to faults in the pincushion correction circuit is the fact that the lines will always be angular. The lines will not run vertical or horizontal across the screen. So, when you come across these odd line symptoms always be suspicious of the pincushion correction circuits. This is the only circuit containing both the vertical and horizontal signals mixed together. This same symptom may also be caused by defective capacitors or a fault in T202, the pincushion correction transformer. This problem has also shown up in other brands of color sets.

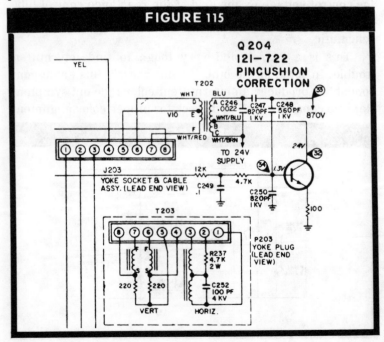

FIGURE 115

ZENITH 14A9C50 CHASSIS
Streaks across screen

137

There were lots of black horizontal lines (approximately ¼ inch apart) over the entire screen of this color set. After considerable troubleshooting, the culprit was found to be an open capacitor (1 mfd, 25 volt NP), C218 from base of the IF AGC transistor (Q202) to ground. This is shown in the schematic of Fig. 116.

This same chassis later on developed a bluish cast over the raster when tuned off channel. This bluish tint would vary with the color level control. Color reception was otherwise normal. This bluish tint was caused by a faulty blanking transistor, Q502.

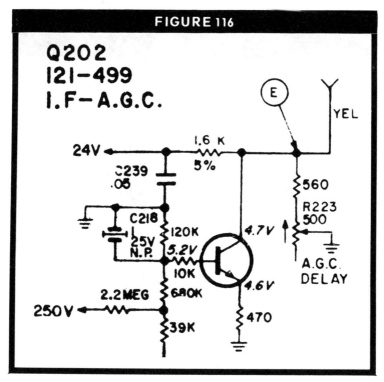

FIGURE 116

138 ZENITH 14A9C50 CHASSIS
No picture or sound

There was no picture or sound on this set and fuse F201 was blown. A few checks revealed that diode CR218 in the power bridge circuit, Fig. 117, was shorted. This diode was replaced and power was applied, but the power supply system was still shorted. A few more checks were made and the short was located in the degaussing coil, L237. These coils will short out to the metal shield around the color picture tube. The degaussing coil can easily be checked out just by unplugging the leads from the chassis. If the short vanishes, the coils are at fault. Replace the coils or check for proper insulation.

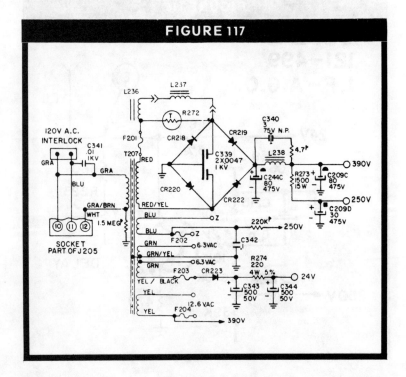

FIGURE 117

ZENITH 14A9C50 CHASSIS USING THE DEMODULATOR IC "CHIP"
Poor color reception

139

The color picture did not look right. Colors were weak or the wrong tint. Also, the hue control had very little affect. The "tip-off" is this: With the color level control turned down, no color, just a black-and-white picture, the CRT screen gray scale will change as the hue control is rotated, usually from a green to a blue tint. The probable cause in a defective IC color demodulator chip, part number 221-39. It has caused this type problem in several chassis. Refer to Fig. 118 for circuit schematic.

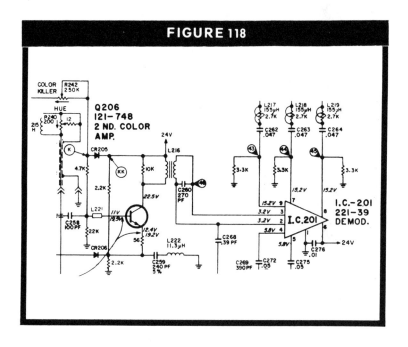

140
ZENITH 14A9C50 CHASSIS
Triple trouble symptoms that may be found in this color TV chassis

A. The screen color shading varied across the CRT from blue on the left side of the raster to black-and-white center and green on the right side. Look for a defective C266 (47 pf) coupling the 3.58-MHz CW from the color oscillator plate to the phase delay line. See the schematic in Fig. 119.

141

B. Picture blurred and milky looking, caused by an intermittent focus condition. Check for a loose black lead (focus connection) at the CRT socket. Replace the socket if defective.

142

C. This set had a blue picture when it was switched off channel and then back on again, but only when a station was broadcasting color information. The condition was due to a floating (open) grid connection on the (B-Y) color difference amplifier, the 6MN8 tube socket.

FIGURE 119

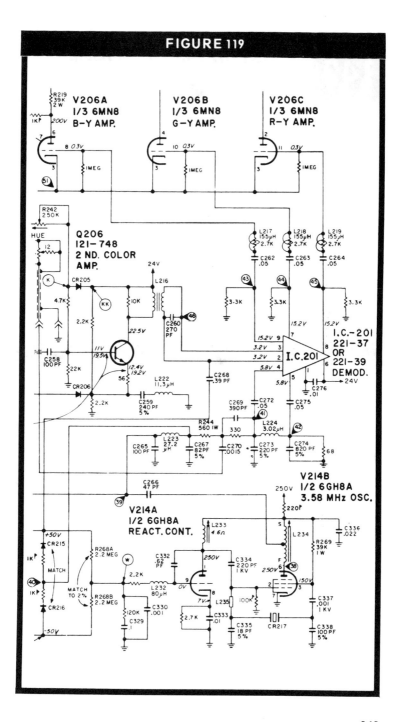

143 ZENITH 14A9C51 CHASSIS
Loss of blue

The raster on this Zenith color TV set had a distinct loss of blue that was of an intermittent nature. The blue G2 control did not seem to have any effect. The plate voltage on the B-Y amplifier (V206A, 6MN8) measured about (40v) and this was too low. Temporarily removing the demodulator IC "chip" caused the blue to return.

This problem was narrowed down to a faulty coupling capacitor, C262, which is a .05 mfd as shown in the schematic of Fig. 120. The capacitor had considerable leakage.

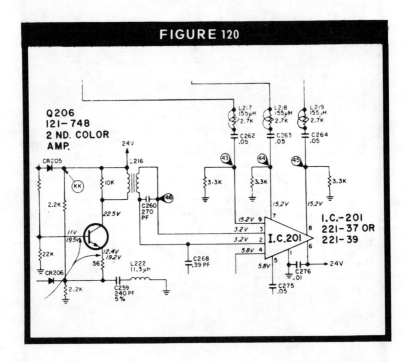
FIGURE 120

ZENITH 14A9C51 CHASSIS
Shimmering picture

Two different component failures in this chassis were found to cause the same picture symptom. The picture had a "pie crust" or "cog wheel" effect that was accompanied by a shimmering picture.

In one chassis this problem was caused by a defective resistor (1000-ohm that had increased in value) at pin 3 of the 6HV5 pulse regulator tube socket. (See the schematic in Fig. 121.) Probable cause for the resistor fault was a shorted 6HV5 tube. Replace the 1000-ohm resistor and the 6HV5 tube. Check the high voltage and adjust if needed.

For the second chassis, which also had the same picture symptom, the problem was traced to a faulty VDR (R255) located in the horizontal sweep output circuit. This is a rare problem.

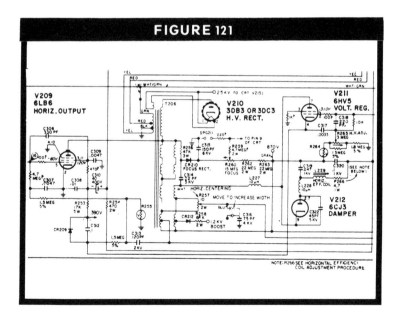

FIGURE 121

145 ZENITH 14A9C51 CHASSIS
Vertical roll

Picture pulls in from the top and bottom and wants to roll vertically. This problem may also be of an intermittent nature. This symptom may also appear in other Zenith color chassis that use this same vertical sweep circuit shown in Fig. 122.

After this set was on a few minutes the vertical oscillator would stop and only a bright vertical line appeared across the screen. The oscilloscope and VTVM were not much help. In fact, when the probes were touched to any part of the vertical oscillator circuit the trouble would clear for a few minutes. A faulty capacitor in the feedback network was suspected, but this was not the case. The defective part was found by substituting resistors in the feedback circuit. The 47K resistor, located between capacitors C237 and C233, had increased in value to 850K and the feedback pulse was too small to sustain oscillation. A very rare and hard to locate malfunction.

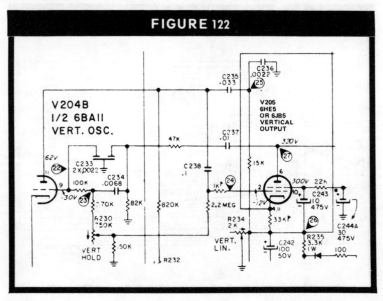

FIGURE 122

ZENITH 14A9C51 CHASSIS
No color

146

This set had a case of no color reception. Test points K and KK were clipped together to open the color amplifiers. The color now "barber-poled," indicating that there was no sync lock. Test point W was grounded (for a zero-beat-check) and the 3.58-MHz color oscillator was tuned to float the color bars across the screen. With test point W ungrounded there was still no color lock action. The color burst amplifier, AFC phase detector, and reactance control circuits were now next on the check list. See the color section schematic in Fig. 123.

The plate voltage of the burst amplifier, V213, was correct. However, the plus or minus 50 volts that should be measured at each end of the matched pair of AFC diodes was low. This was about plus or minus 25 volts. The primary winding coil, L230, was open. The plate voltage checked out correct because of the 8.2K resistor across the primary winding. After a new L230 coil was installed the color picture locked in solid. Use an oscilloscope to troubleshoot this type of color sync problem.

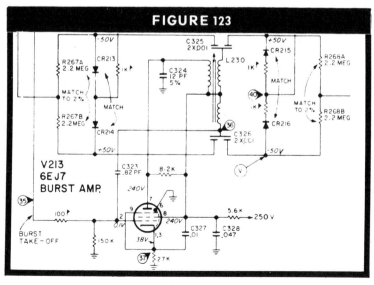

FIGURE 123

147 ZENITH 14A10C19 CHASSIS
Brightness level varies

The picture on this TV receiver displayed an intermittent "pulsing" of the brightness level when a TV station was being received. However, when switched to an inactive channel the brightness would be steady.

The defect was found to be an intermittently open AGC cathode by-pass capacitor (4 mfd, 150v DC designated as C234 in the schematic of Fig. 124. This capacitor, when defective, has been known to cause many odd AGC problems on other Zenith color chassis. This same capacitor fault will also cause improper AGC tracking when switched from a strong TV station signal to a weaker one. Replace C234 and readjust the AGC level.

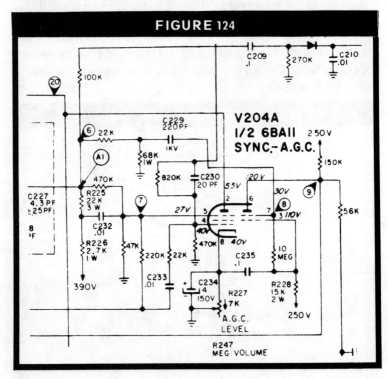

FIGURE 124

ZENITH 14A10C27Z CHASSIS
High line voltage areas

148

It has been found that at high line voltages (130v), the ½-watt 470K resistor in the screen grid circuit of the 6LB6 may overheat and possibly open. See Fig. 125. This will cause a loss of HV and raster and open the fuse in the cathode of the 6LB6.

Service Tip: Check the connection of the 470K resistor to the 390v B+ line. The above condition may exist if the 470K resistor is connected to the boost voltage terminal instead of 390 volts. Change the connection if not as indicated. If the condition exists, also check R263 (17K) and the 470-ohm resistor in the screen circuit.

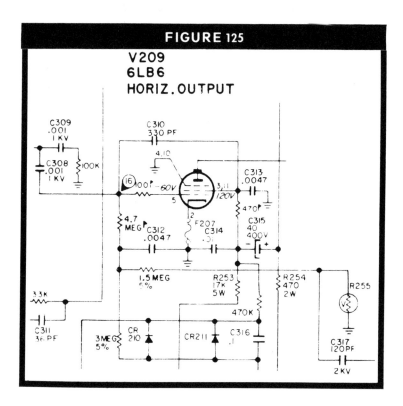

149

ZENITH 12A12C52 CHASSIS
No sync and loss of color

The picture on this color set displayed a loss of picture sync locking action with only "flashes" of color showing up in the picture intermittently. This behavior was puzzling to say the least, so we delved right into the sync and AGC circuits as shown in the schematic of Fig. 126.

All of these troubles were caused by a defective (open) capacitor C243 (180 pf) in the sync and AGC stage. Also check diode CR201, as it will cause this same picture trouble symptom should it become defective. Replace C243 or diode CR201 and readjust the AGC and AGC delay controls as required.

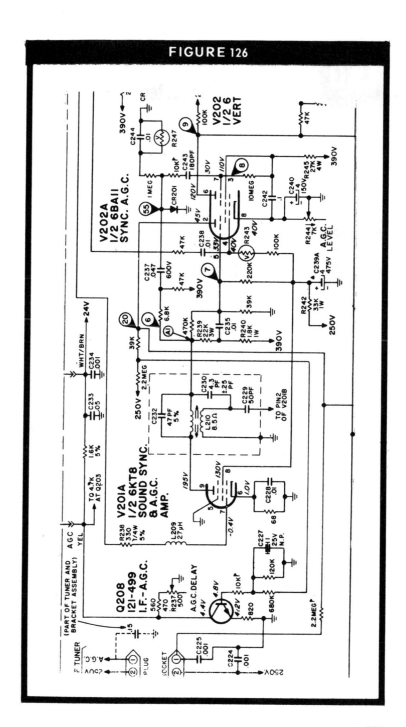

FIGURE 126

150
ZENITH 12A12C52 CHASSIS
No raster

Probable Cause: Open third video amplifier transistor (121-746). Refer to Fig. 127 for this video amplifier circuit. This transistor fault will simulate a "blanking" condition.

151
ZENITH 12A12C52 CHASSIS
Loss of high voltage

Possible Cause: A shorted third video amplifier transistor (121-746) which will reduce the CRT cathode-grid bias.
    Note: In the two above case-histories the "set-up" line could be "normally" adjusted with the G2 controls, indicating a "Y" channel malfunction. See Fig. 127.

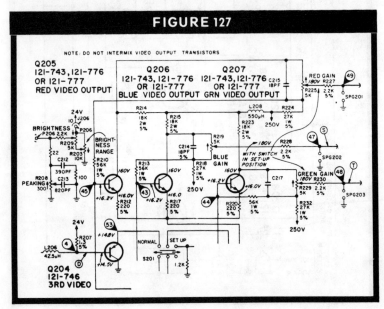

ZENITH 12A12C52 CHASSIS   152

This color set had a picture that was not clear and had a double image blue ghost appearance. With a crosshatch generator hooked-up, it was determined that the blue convergence was way out. And no blue horizontal line dynamic control action was noted. Check diode CR601D. It may have shorted and burned up control R609 and coil L601.

Replace the diode, the complete unit, grid coil L601 and R609 as shown in Fig. 128. Also install a 47-ohm 2-watt resistor in series with the 180-ohm control (R609).

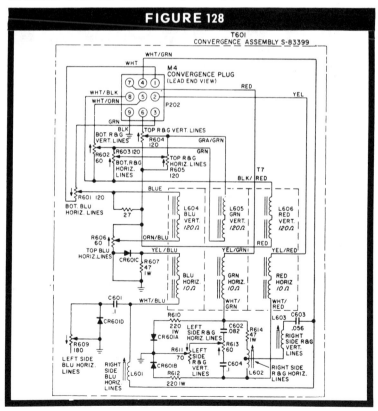

153 ZENITH 12A12C52 CHASSIS

At times the picture on this set would be too bright. Along with this intermittent brightness condition, it was noted that the set-up line could not be obtained. These were the tip-off symptoms.

The cause of this trouble was a shorted capacitor. This is C218 (see Fig. 129), a .01 mfd capacitor in the control grid circuit of V213, the color CRT. Also, check or replace spark gap SPG 211.

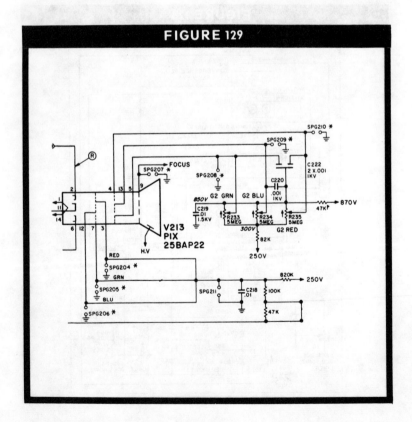

FIGURE 129

ZENITH 12A12C52, 20X1C36 AND OTHER COLOR CHASSIS
Color picture moire

In some instances, and under certain conditions, an effect called "moire" may be exhibited on the screen (picture) of a color TV receiver. This will show up as a group of narrow, usually swirling (but stationary) lines which can appear anywhere in the picture in varying degrees of intensity. It may appear somewhat similar to an enlarged "thumb print."

Several factors can contribute to this condition. Such as, beam size, type of scene, and in particular, the degree of focus.

Moire can be eliminated, or reduced to an acceptable degree by making a simple circuit change as will be shown in the following chassis groups. Essentially, a slight reduction in the cut off voltage of the CRT is achieved.

Color difference systems with a 150v CRT cut off voltage, including the following chassis: 20X1C36, 20X1C38, 20Y1C38, 20Y1C48, 20Y1C50, 14Z8C50 and 14A9C51.

As shown in the schematic of Fig. 130A, change the 82K, 2w to a 47K, 3w resistor and this will reduce cut off by 20 percent (120v).

Color difference systems with a 140v (red) and 150v (green and blue) CRT cut off, covering the following chassis: 16Z7C50, 16Z7C50Z and 16Z8C50.

As shown in the schematic of Fig. 130B, changing the 39K, 4w to a 33K, 3w resistor will reduce the CRT cut off by 20 percent (112v and 120v).

The RGB system with 150v CRT cut off, covering the 12A12C52 chassis. Paralleling the 820K, ½w with another 820K, ½w resistor will reduce cut off by 20 percent (120v) as shown in Fig. 130C. Removing jumper wire across the 47K resistor will reduce cut-off by about 10 volts.

FIGURE 130

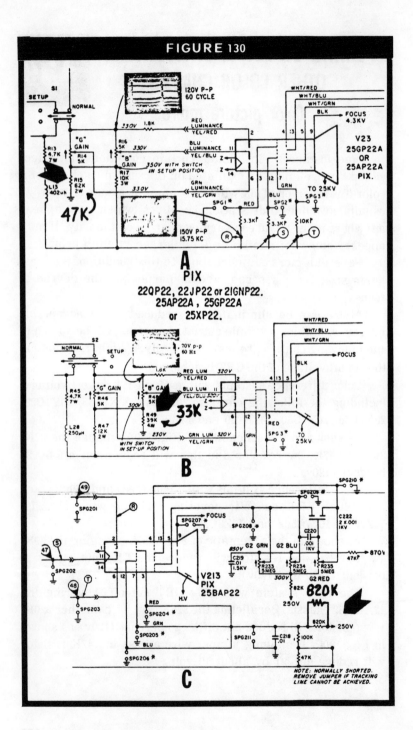

ZENITH 4B25C19 CHASSIS
Picture "ghost"

There were "ghosts" or shadows on the pictures received by this color set.

The trouble appeared to be due to an antenna fault. Actually, the trouble was a faulty transistor, 121-746, third video amplifier shown in Fig. 131. By accidentally "healing" the transistor, the brightness increased excessively and AGC action was also affected. Retracking and readjusting the AGC resulted in a normal picture without "ghosts." After a little circuit coolant was applied to the transistor, it would break down again. Readjusting the AGC again caused the ghosts to reappear. A new transistor and alignment solved this spook problem.

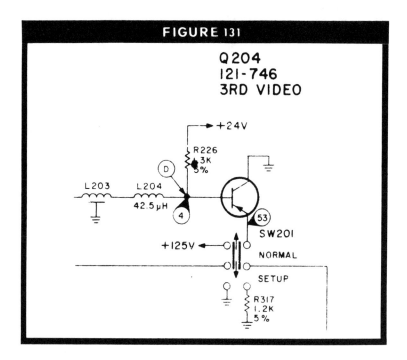

156 ZENITH 4B25C19 CHASSIS
Loss of color sync

There was no color lock on this color TV chassis. The color was out-of-sync and had a barber pole appearance. When the normal-align switch was placed in the align position, the color was lost completely. The trouble was traced to a faulty (shorted) capacitor (C1011, .05 mfd) shown in the schematic of Fig. 132. Also, capacitor C1009, .05 mfd, will cause the same trouble. These components are located on the 9-27 module.

157 ZENITH 4B25C19 CHASSIS
No raster and-or loss of brightness

This set had no brightness. The problem turned out to be in the color section. IC 902, located on the 9-37 chroma module assembly, was found to be defective. See Fig. 132.

Another chassis was found with no raster and an open fuse in the horizontal sweep output stage. This trouble was caused by shorted turns in the horizontal windings of the pincushion saturable reactor.

FIGURE 132

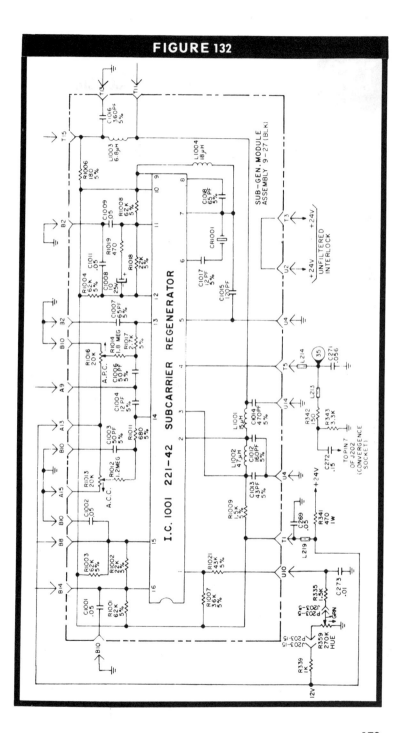

158 ZENITH 4B25C19 CHASSIS
Screen dim or black-out

It was noted when the picture went dim that the screen controls and brightness limiter control would brighten the screen up, but the picture had a negative appearance. However, the "tip-off" came when we noticed that the color was missing.

For a fast check, the chroma module assembly was changed (Part Number 9-34 orange color) and the picture and color reception returned to normal. Refer to Fig. 133 for this module circuit. The IC "chips" on this module were changed and IC 221-46 (chroma demodulator) was found to be defective. After the "chip" is replaced, make an APC color alignment and cross-talk adjustment.

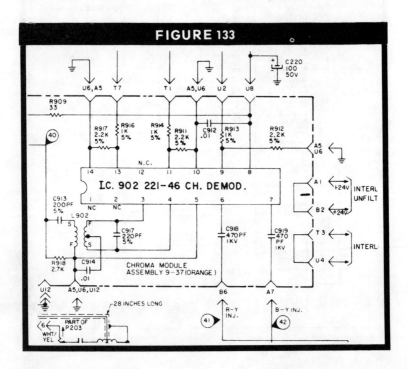

FIGURE 133

ZENITH 4B25C19 CHASSIS
Intermittent sound condition

159

Sound went off after set warmed up, or it was intermittent. Check the sound module. Part Number 150-206. Both conditions are caused by a defective transformer T102 or poor solder connections on the circuit board shown in Fig. 134. Also the sound "chip" (IC1101) may be defective. Repair the circuit board or replace the faulty components. As a last resort, replace the sound module assembly. However, don't overlook Q103, the sound detector amplifier found in the IF sub-assembly. Also, check the plug and cabling connections.

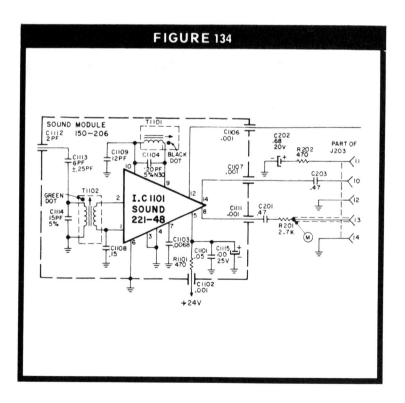

FIGURE 134

160 ZENITH 4B25C19 CHASSIS
Blue missing in the picture

There wasn't a trace of blue in the color picture. Voltages at cathodes of the CRT were found to be correct. (See the schematic in Fig. 135.) The CRT grid voltages were normal, as were the C2 voltages. Voltages at the three video output transistors were normal, too. A very dim blue line could be achieved with switch in the setup position and the blue G2 control at maximum. Voltage at the center arm of the blue gain control read 20 volts lower than the other two.

This trouble was caused by a defective (open) resistor, R292 (2.2K) in the blue video output stage. Replace the resistor and perform gray-scale tracking adjustments if needed.

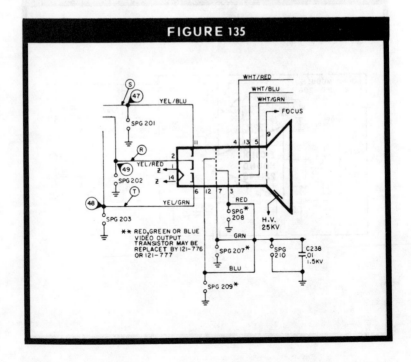

FIGURE 135

ZENITH 4B25C19 CHASSIS
Screen too bright

161

The picture on this color TV set had far too much brightness and would also blur out of focus. The brightness control had very little effect. The (.4 amp) fuse may also blow.

In this chassis the fault was a defective picture peak control (R244) shown in Fig. 136. The control had developed leakage to the case. The resistance from high side to ground read approximately 150 ohms at certain settings of the control.

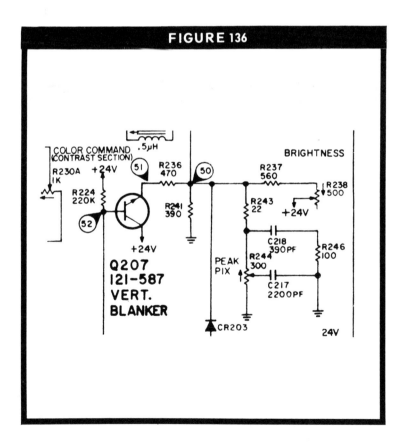

FIGURE 136

162

ZENITH 12B14C50 CHASSIS
No raster

The sound was fine on this chassis, but there was no screen brightness, just a dark picture tube. It was noted that a blue line varying in intensity was displayed when the set-up switch was placed in the set-up position. The high-voltage was correct at the CRT anode.

A few checks revealed that the 3rd video amplifier transistor was defective and caused the loss of raster. However, a faulty blue video output transistor had caused the blue line intensity variation. Fig. 137 is a schematic of these video circuits.

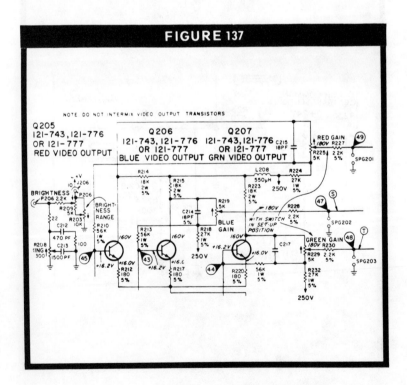

ZENITH 12B14C52 CHASSIS
Very weak color

This set appeared to have no color, but a close look revealed a very weak, washed out picture. All voltages in the color section checked very close to normal. The trouble was found by signal tracing the chroma waveforms with an oscilloscope.

The trouble was found, and this can be a tough one, to be an open winding in the first color amplifier transformer (L213 plate winding). The plate voltage, pin 9, of this stage checked about normal, because of the 10K resistor from the screen grid, pin 8, of this circuit. However, the scope caught the loss of color gain for this stage in short order. Replace the transformer and align the color amplifier stage. The circuit is shown in Fig. 138.

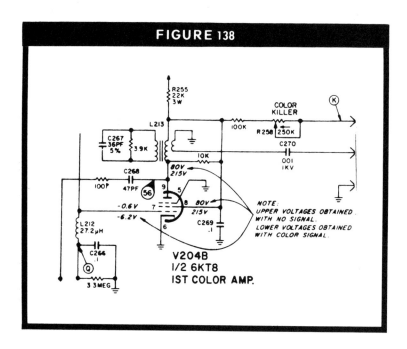

FIGURE 138

164

ZENITH 12B14C52 CHASSIS
Low level color

The black-and-white picture on this set was good and the sound was normal, but the color information was very weak. The color amplifiers checked out good and the color demodulator chip IC701 was changed, but the weak color was still evident.

After checking the 3.58 MHz oscillator circuit (shown in Fig. 139) with an oscilloscope, the trouble was narrowed down to resistor R280 (560 ohms) which had increased in value to 1.5K and weakened CW drive signal to the chroma demodulator "chip." This resistor is located in the coupling circuit between the 3.58 MHz oscillator output and input to the demodulator.

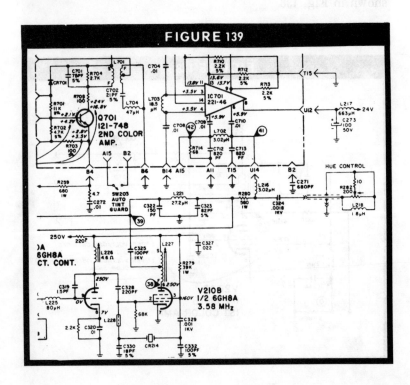

ZENITH 12B14C52 CHASSIS
Low volume and distorted audio

165

This color chassis was producing an excellent picture, but the sound level was low and a slight distortion was noted. New tubes in the audio stages did not help. All voltages in the sound section were measured, but were found to be normal. (See Fig. 140.) An audio signal tracer was used and all of the stages appeared to be working properly. On a "hunch," the audio output transformer (T206) was replaced by a new one and the sound came on loud and clear.

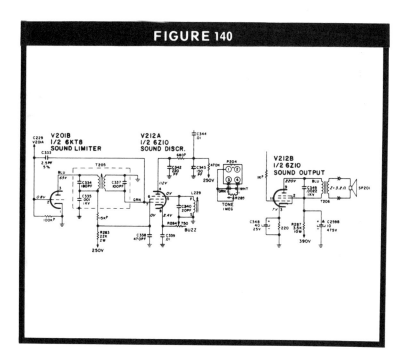

FIGURE 140

166

ZENITH 12B14C51 CHASSIS
A blue blood color receiver

The picture on this set was good except that blue bled out from the picture details on the left side of the screen. A cross-hatch generator showed very poor blue convergence on the left side of the screen. Some touch-up convergence adjustments did not help. However, the left blue horizontal lines control had no affect on the pattern, so this was a good clue. We took a good look at the convergence assembly panel. It was found that this control, R609 (see schematic in Fig. 141), had opened. A few more checks were made to find out why R609 went up in smoke. The trouble actually started when diode CR601D shorted out. There are four diodes in this unit, so replace the complete package and make a convergence check.

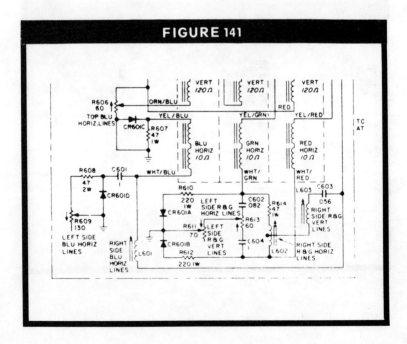

FIGURE 141

ZENITH 12B14C51 CHASSIS
Washed out color

167

This Zenith color set had a very weak, washed out color picture. It was so weak that the complaint was no color at all. Later, on the service bench, it was noted that all voltages in the color section checked out very close to normal readings. This could have been a tough one to find, but the oscilloscope made it easy. The problem was located in the first color amplifier transformer, L213, shown in Fig. 142. The secondary plate coil winding was open. The plate voltage, pin 9, of the first color stage was about normal because of the 10K resistor feeding voltage from the screen grid, pin 8, of this stage. However, the oscilloscope caught the loss of color gain in this stage very rapidly. After the transformer was replaced and color alignment performed, bright color once again appeared on the screen.

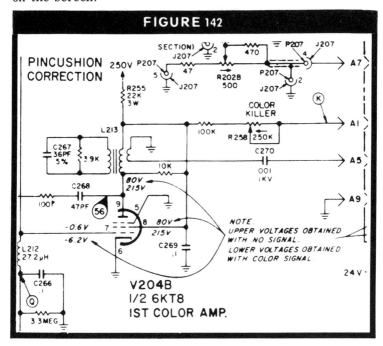

168
ZENITH 12B14C51 CHASSIS
Color trouble

This set had color that would come and go or would go entirely. With test point W shorted to ground and test points K and KK connected together, the 3.58-MHz color oscillator zero beat properly. However, when these test point shorts were removed, the 3.58 MHz color oscillator was pulled way off frequency. The voltages measured at the AFC diodes were +51 and -28 volts. These should balance at plus or minus 50 volts.

The problem was found to be defective AFC color phase detector diodes, a matched pair of diodes shown as CR212 and CR213 in the schematic of Fig. 143. Replace these diodes and zero beat the 3.58-MHz oscillator.

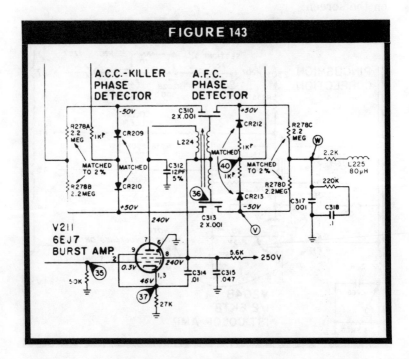
FIGURE 143

ZENITH 14B9C50 CHASSIS
Bars across the screen

169

These may appear as dark or light gray bars across the screen. They usually have a jail bar effect, and usually will widen or narrow when adjusting the fine tuning, color level, hue or color killer controls.

There is a possibility that a diode or a transistor may have been inadvertently reversed during installation. Always observe the polarities carefully. In this chassis, diode CR206 (see Fig. 144), located at the 2nd color amplifier stage, was installed backwards. This same type of jail bar effect may also be caused by a defective blanker diode or transistor such as CR204 or Q205.

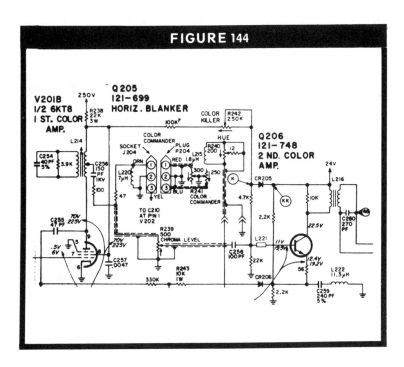

FIGURE 144

170 ZENITH 14CC14 CHASSIS
No color

This small portable set had a sharp black and white picture but not a trace of color. For a problem of this nature you should defeat the color killer stage if the set has one and then measure the automatic color control voltage (ACC). In this chassis the ACC voltage at test point A measured .50 volt and this was cutting off the 1st color amplifier stage. Diode CR210 (Fig. 145) was found to be faulty (open) and was causing incorrect ACC voltage at the color amplifier grid. If diode CR200 opens, it will also cause a no color condition. And if diodes CR201-CR202 were faulty, you would not have any blanking to the "chip." These diodes also eliminate any ripple voltage between blanking pulses that are used to provide a low amplitude pulse for cutting off the color demodulator chip (IC201) during the horizontal blanking interval. Leakage in diode CR210 may cause strong color with no ACC action at all.

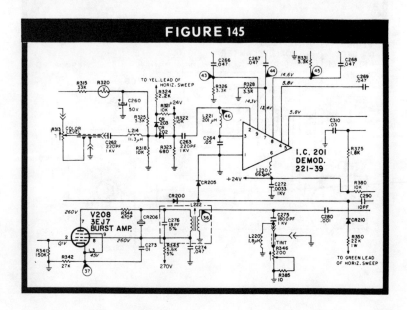

FIGURE 145

ZENITH 25NC37 CHASSIS
Silicon rectifier failures

171

These Zenith color chassis (and many other color TV brands) have been found with shorted silicon rectifiers. This occurs on later model color sets with automatic degaussing.

It seems that the rubber HV anode cup deteriorates or moisture forms under it and a HV arc occurs. On some of the Zenith 25NC37 chassis, the degaussing coils are located very close to the HV anode button on the CRT. When a HV arc occurs, this sends a tremendous spike or pulse of voltage through the degaussing coils and thus onto the silicon rectifiers SE1 and SE2 (see Fig. 146) and, of course, blows out one or more silicon diodes.

Reposition the degaussing coils as far from the HV anode cap as possible and replace the HV anode cup with the small size rubber cup. Clean the area very thoroughly around the CRT anode HV connection. Make sure the HV shunt regulation system is working properly. Check and adjust the HV according to the manufacturers specifications. This should eliminate "shorted diode failure" call backs. However, spike voltage transients on the incoming power line to the receiver may also cause this same type of silicon diode failure.

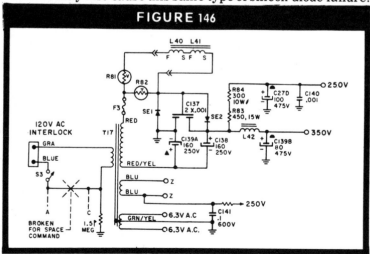

172

ZENITH 20X1C36 COLOR CHASSIS
A short circuit problem

This color set worked fine until the hue control was turned counterclockwise, then the B+ fuse (F1) would blow. You just don't run across this type of trouble every day.

The trouble was located and a look at the color circuit in Fig. 147 will show how this can happen. Capacitor C138 located between the phase detector coil (L38) and the hue control becomes shorted and puts B+ to ground through the 5.6K screen grid resistor of the burst amplifier stage V16, thus blowing the fuse. And this will also burn up the hue control and the 5.6K resistor. Replace C138, a .01 mfd capacitor, the 5.6K resistor and the hue control. Then check for proper hue control tint range and adjust if needed.

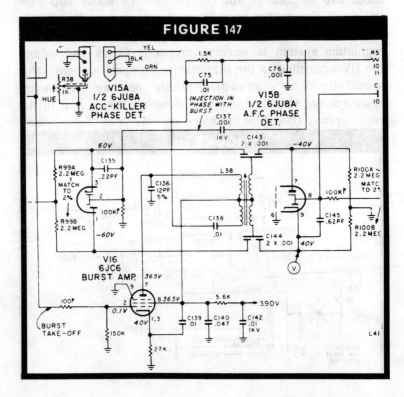

FIGURE 147

ZENITH 20X1C38 CHASSIS
Color sync trouble

173

This set displayed a color picture sync problem. The voltage checked at test point "W" would only swing plus and minus one volt. An oscilloscope check revealed a faulty gating pulse at the burst amplifier plate circuit. The color burst and video were present at the grid.

The trouble proved to be a faulty 47 pf capacitor coupling the chroma signal to the burst amplifier grid. (See the schematic in Fig. 148.) As a result the color burst amplifier stage conducted continuously. A new capacitor gave us good color picture lock.

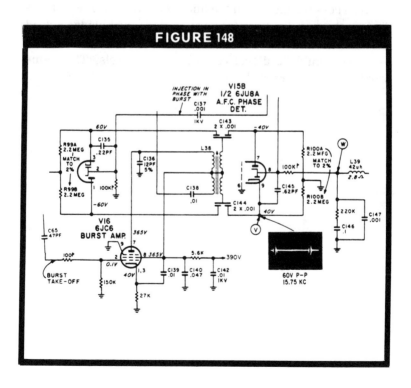

ZENITH 15Y6C16 CHASSIS
Convergence trouble

If the red and green horizontal lines on the left side of screen will not converge, check with an oscilloscope for proper waveforms coming into the convergence panel, see Fig. 149. If the waveform is correct, but you have a bow in the horizontal lines, check the clamping diodes.

This chassis had a defective X9B clamping diode. These diodes will short, open or develop leakage.

The diodes are used to "clamp" the dynamic convergence sweep at the screen center, thus giving the beams more effect at the ends of each sweep. A defective diode will cause the line to bow and the controls will have no response.

Short out the diode and the line should bow; if it won't, the diode is shorted. To check for an open diode, just bridge a good one across it and look for results. Of course, other troubles on the panel could be defective coils and controls. This same problem has occurred in many other makes of color TV sets.

FIGURE 149

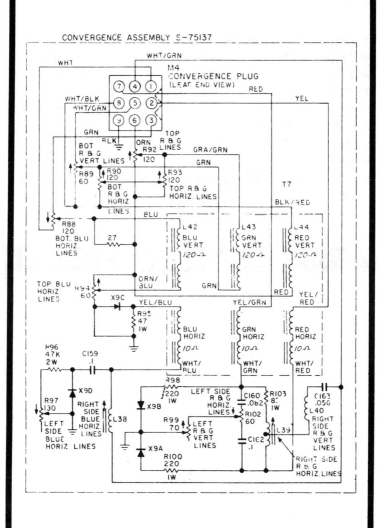

175

ZENITH 20Y1C38 CHASSIS
Hue control varies color

The hue control varied the color level and had very little affect on the phase variance. This can be a tough one to troubleshoot because the defective component may check out good. Also, the voltage at the control grid of the 1st color amplifier stage has a 3-volt range variance as the hue control is rotated through its full range.

Capacitor C156, a .01 mfd unit (Fig. 150) in series with the arm of hue control R43, had developed leakage. Replace C156 and touch-up the AFC phase detector transformer, L41, if needed.

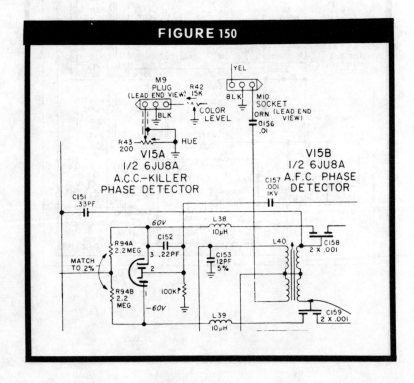

FIGURE 150

ZENITH 20Y1C48 CHASSIS
Picture convergence drift

176

An in and out of convergence drift problem found in this chassis was a "rare" trouble. When the channel selector was turned off station, then back on, or even during a sudden picture change, it looked like the convergence dropped out. But then, after about 5 or 10 seconds, the picture would slowly drift back into proper convergence. This symptom was very pronounced with a cross-hatch pattern on the screen.

This "video information transient" symptom is very likely caused by a faulty electrolytic capacitor. The most likely faulty ones (intermittently open with a high power factor) are the large value electrolytic capacitors in the vertical sweep or convergence pulse forming circuits. The component which caused this odd drift problem in this set was the large 100 mfd electrolytic unit in the cathode circuit of V10, vertical output stage, which is shown as C62 in the schematic of Fig. 151. This capacitor is not only a cathode bypass, but it also shapes the vertical dynamic convergence pulses that are fed into the convergence control panel. Replace C62 and then recheck the overall convergence.

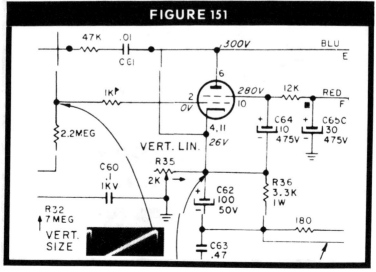

FIGURE 151

ZENITH 20Y1C50 CHASSIS
Weak color

The picture on this color receiver displayed weak "reds" during color program reception. In fact, the overall color information appeared to be weak. This weak color threw us a curve as we went right to checking the chroma amplifier stages. However, the trouble was found in the R-Y demodulator stages shown in the schematic of Fig. 152. A faulty resistor (680 ohms) in the cathode circuit of V13 (R-Y demodulator) had increased in value to 3000 ohms.

Service Note: Weak color(s) may also be caused by an open peaking coil(s) L19, L20 or L22 in the output circuit of the demodulators.

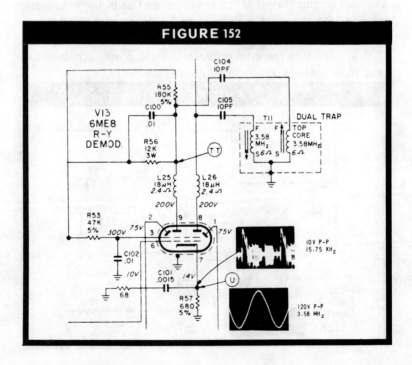

ZENITH 14Z8C50 CHASSIS HV Regulation trouble — 178

This set had an out-of-focus picture that also looked a little washed-out, too. Frequent "snapping and popping" was noted, particularly when the program changed from color to black-and-white information. The HV control had no effect. The HV was measured and was found to be 32,000 volts.

The pulse HV regulator stage (V11) was checked out and R101 was found to be open. This is a 220 ohm resistor in the cathode of the 6HV5 regulator tube. Also check capacitor C181 and install a new 6HV5 tube. This tube shorts out and burns out R101. Now adjust the HV for the correct value. (See Fig. 153)

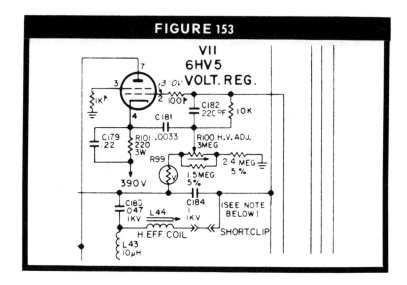

179

ZENITH 14X8C50 CHASSIS
Left side of raster green

This color receiver had a very strange symptom. With the set-up switch in the set-up position, the green line appeared twice as wide at the left side of the raster. The green line then narrowed toward the center and faded out about 4 inches from the right side of the screen. The red and blue lines appeared normal.

The possible cause: A faulty (shorted) capacitor C131 (.05 mfd) in the control grid circuit of the 6MN8 G-Y amplifier stage. This circuit is shown in Fig. 154.

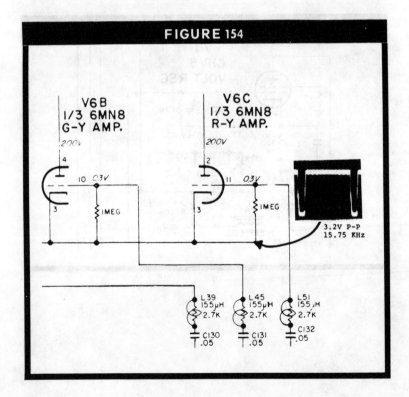

ZENITH 14Z8C50 CHASSIS
Dark horizontal hum bar

180

This color chassis, at first glance, looked like it had an open filter capacitor in the B+ power supply section. A scope was used to check the B+ supply, but all was found in good order. The screen produced a black horizontal hum bar approximately 4 inches wide that moved vertically through the raster. By using the oscilloscope, the trouble was tracked down in the color section. A defective (open) screen bypass capacitor, C125 (.0047 mfd), was located at V1B (Fig. 155) in the first color amplifier stage.

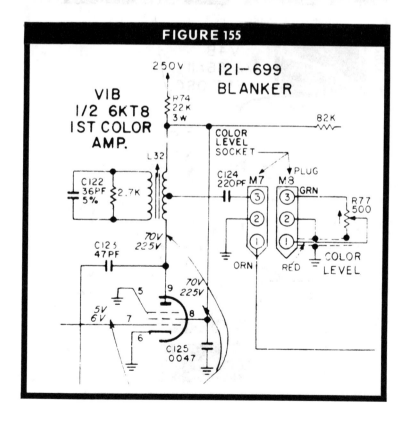

FIGURE 155

181 ZENITH 14Z8C50 CHASSIS
Vertical roll

The picture did a slow vertical roll and lacked vertical locking action. Voltage measurement at the vertical blanker transistor revealed a positive voltage at the base and emitter.

The probable cause was a defective diode in the vertical oscillator stage. A resistance check of diode X8 revealed it was shorted. This same problem has been noted on other Zenith chassis (black-and-white or color) that use this diode in the cathode circuit of the vertical oscillator. Refer to Fig. 156 for this vertical oscillator circuit schematic.

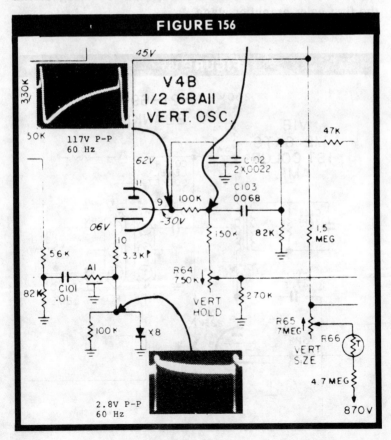

FIGURE 156

ZENITH 14Z8C50 CHASSIS
Tracking modification

182

Improved IF-RF AGC tracking modification (this circuit change is for all Zenith color TV chassis with transistorized IF systems).

Later production color receivers utilizing the solid-state IF system now incorporates a circuit modification providing improved tracking of IF-RF AGC voltages under all signal conditions.

Essentially, the B+ distribution to the AGC amplifier tube has been revised. A 39K resistor has been added in series with the IF-AGC delay circuit and the 2.2 megohm resistor coupled from B+ (250v) to the AGC tube plate has been relocated. (Fig. 157)

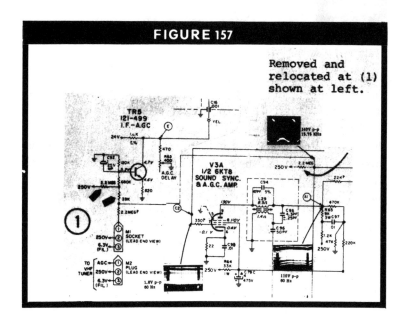

FIGURE 157

183

ZENITH 16Z7C19 CHASSIS
Wrong colors

Erroneous colors and a "shifting" hue were the complaints given for this Zenith color receiver. This type trouble can be hard to find and check out.

The trouble on this chassis was caused by a defective capacitor (C205, 1.5 pf) located between the plate and grid of the 6GH8, 3.58 MHz reactance control circuit as shown in Fig. 158. This capacitor is too small a value to check, so component substitution was used to solve this trouble.

184

ZENITH 16Z7C19 CHASSIS
Color comes and goes

This set had an intermittent color reception problem and as you well know this type of trouble can be hard to solve.

This trouble was found to be caused by a defective phase shift capacitor (C145, 220 pf) found in the 3.58 MHz transmission line between the 6BV11 demodulator injection grids. This is also shown in the schematic of Fig. 158. Check and touch-up color alignment if needed after C145 is replaced.

ZENITH 16Z7C19 CHASSIS
Colors not correct

185

A color-bar generator was connected to this set and it was noted the colors had a 90 degree phase shift. A color alignment was made, but did not cure the phase shift. Service tip: The defective component was actually found while we were performing the color alignment.

This phase shift condition was caused by a defective third color amplifier plate coil, shown as L34 in the schematic of Fig. 158. After this coil is replaced, check the set for proper color alignment.

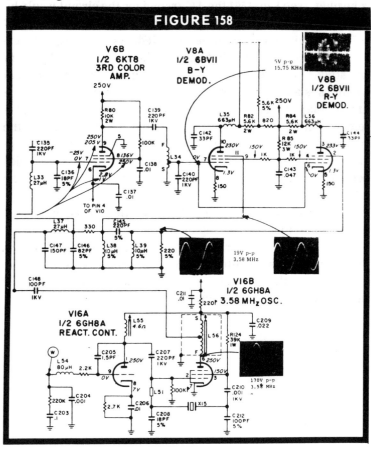

186 ZENITH 16Z7C50Z CHASSIS
Ringing

This rare malfunction in the deflection circuit of a color TV receiver can cause ringing. Raster lines which appear as sine waves (approximately 1/8" in height) for the first few inches of horizontal scan may not be due to a fault in the horizontal sweep circuits as one might suspect.

For example, should the center tap of the pincushion transformer open (i.e., broken connection at the deflection yoke plug, (see Fig. 159) the AC ground for the circuit would be removed. Also, the two damping resistors across the vertical deflection windings would be removed. As a result, the vertical and horizontal windings would be more subject to crosstalk. The natural resonant frequency of the vertical windings is approximately 240 kHz. During horizontal retrace time (collapsing of the sweep voltage), considerable energy is released in the horizontal windings and it is coupled into the vertical windings. This can set up a ringing action at approximately 240 kHz. The ringing decays after a few cycles because it is not sustained immediately following the horizontal retrace pulse.

This ringing, appearing in the vertical yoke windings, exhibits itself as severe ringing on the left side of the screen. Thus, the raster lines for the first two inches of scan appear as sine waves. This condition is most apparent when the set-up switch is in the "set-up" position. These yoke plugs will develop poor pin connections.

FIGURE 159

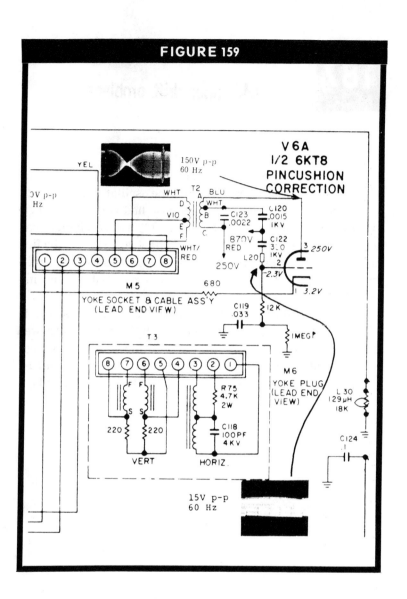

187

ZENITH 16Z7C50 CHASSIS
VHF tuner drift problem

When this color receiver was first turned on, the VHF stations could not be fine tuned properly and the oscillator appeared to be drifting. The fine tuning control also had very little effect. The AFC control voltage was removed, but no change was noted. After the set would warm up (about 20 minutes) the fine tuning then performed normally. The two tubes were changed and then a good tuner cleaning was performed, but no improvement was noted.

The schematic for this type 175-595 VHF tuner is shown in Fig. 160. After a few more checks, the problem was narrowed down to an intermittent varactor AFC diode (Part Number 103-146) in the tuner oscillator stage. This is diode X1 in Fig. 160. An ohmmeter check revealed "infinity" readings in both directions.

FIGURE 160

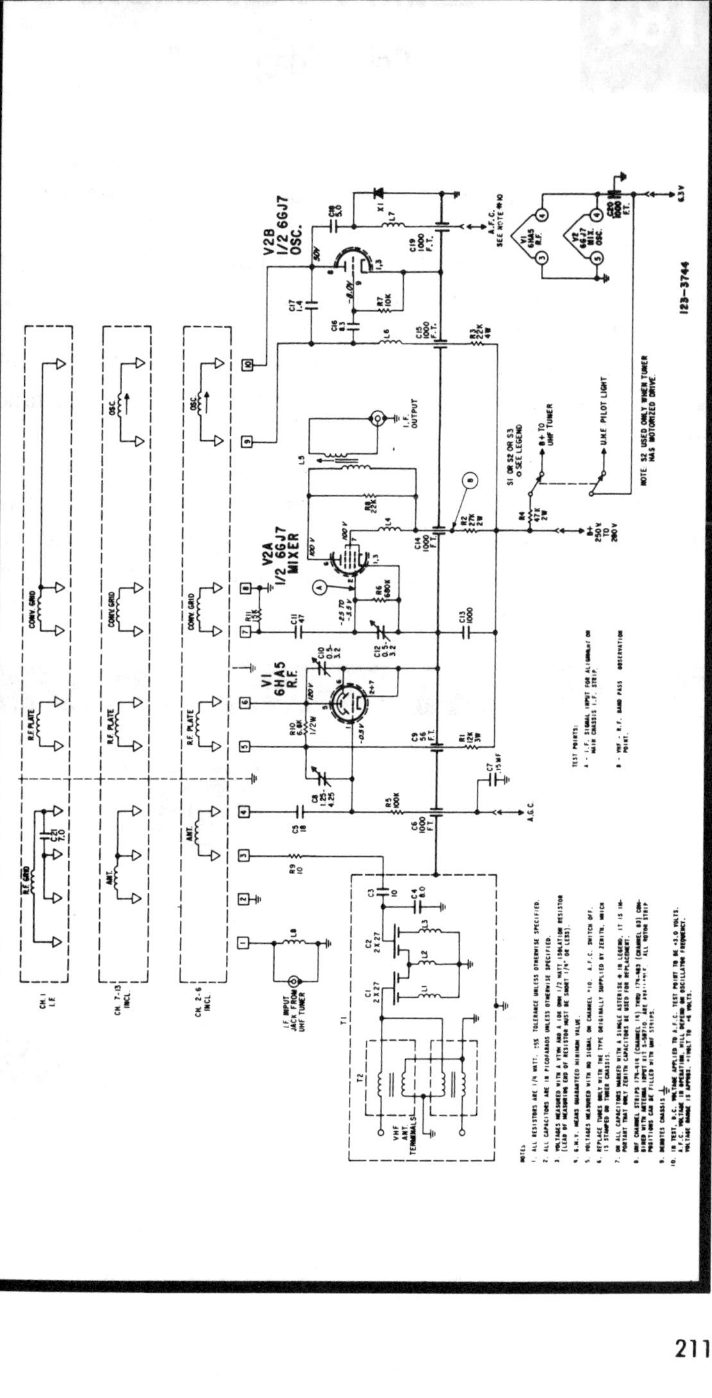

188

ZENITH 16Z7C50 CHASSIS
Color "drop out"

In this receiver, the color dropped out and horizontal color lines (rainbow) appeared instead of a normal color picture. The adjustment of coil L55 (color oscillator frequency adjustment) will be difficult. When test point "W" is shorted to ground for the adjustment procedure, a "zero beat" cannot be achieved.

Reverse the series components, L54 and the 2.2K resistor, between test point "W" and control grid of V16A, reactance control tube.

Note: On some chassis the coil (L54) will be connected to grid of V16A. Coil L54 should be connected to test point "W" and the 2.2K resistor goes to the control grid. The late production chassis have this change which is correctly shown in the circuit of Fig. 161.

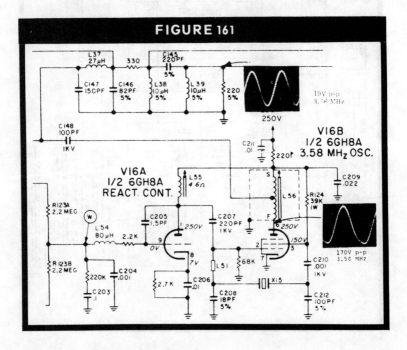

ZENITH 16Z7C50Z CHASSIS
Not enough vertical sweep

189

The sound was normal on this set, but only a narrow, white horizontal band across the center of the screen was visible, thus indicating very little vertical deflection. The tubes were good and the vertical height and linearity controls had no affect on the vertical deflection. We first suspected a faulty cathode capacitor, C113, see schematic in Fig. 162. However, even though this capacitor has been known to be a trouble maker, it was found to be good in this chassis. The vertical oscillator was also operating properly. The trouble was located as an open 12K screen control resistor located at pin 10 of the vertical output tube, V5. Replace the 6HE5 tube, since it probably shorted and caused the resistor to fail. Also check capacitors C115 and C116B. This is a rare vertical deflection problem and can be easily overlooked.

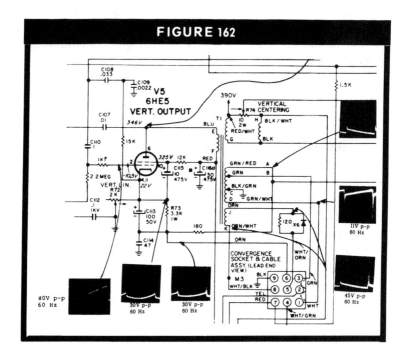

FIGURE 162

190 ZENITH 16Z7C60Z CHASSIS
Erratic AGC

This Zenith color set had an unusual situation where a tuner B+ malfunction was disguised by erratic AGC action.

The case involved a set with a 16Z7C50Z chassis which was exhibiting the following symptoms. See Fig. 163 for the tuner schematic.

The AGC control was very "sensitive" and adjustment was exceptionally critical. Strong signals tended to overload while weak signals exhibited an abnormally "snowy" picture. Certain settings of the AGC control resulted in brightness and signal flickering in the picture. The AGC voltage at both the tuner and IF stages was erratic, as well as the composite video information at the video detectors (test points C1 and C2). Although the condition appeared as an AGC malfunction, later checks proved otherwise.

The possibility of a defective tuner was quickly eliminated by substitution, but to no avail.

Subsequent checks revealed that a fixed DC voltage (substituting for AGC voltage) applied to either the IF and-or RF stages tended to stabilize the picture.

Finally, by tuner voltage checks, an open 9.1K resistor, R1 in the B+ circuit of the RF amplifier (6HA5), was discovered. No other components were defective. Since this resistor is external to the tuner, the 6HA5 plate voltage on the substitute tuner was also absent.

The resistor was replaced, and the "erratic" AGC action was remedied.

FIGURE 163

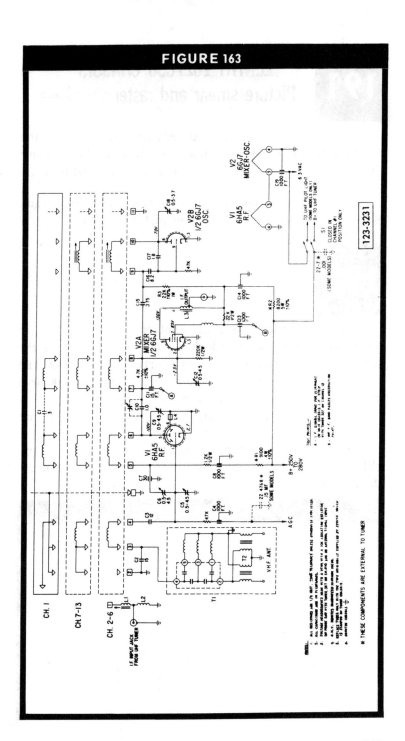

191
ZENITH 16Z7C50 CHASSIS
Picture smear and raster shrinkage

Excessive picture smear or a blurred rater pulled in from the sides are double troubles that may crop-up in this Zenith color chassis. The same component failure will cause both symptoms.

Problem number one: Excessive picture smear (color and black-and-white). A lack or loss of picture definition.

Problem number two: On another chassis the raster pulled in approximately 3 inches from the left side of the screen. The picture was also very blurred. The contrast and brightness controls had little or no effect. Very weak or a complete loss of video information was also evident.

The fault was found in the video "Y" amplifier circuit shown in Fig. 164. A defective (shorted or leaky) coupling capacitor, C76 a .1 mfd, at the control grid of the "Y" amplifier. This may also be an intermittent condition just to make things interesting.

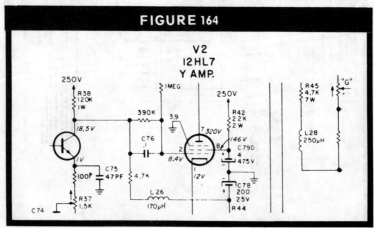

ZENITH 16Z8C19 CHASSIS
Picture interference patterns
193

When confronted with a weird looking pattern on the TV screen, don't overlook the dynamic pincushion circuits as a probable cause. Slight component variations and malfunctions in these circuits may cause some unusual or misleading patterns. The Zenith 16Z8C19 pincushion circuit shown in Fig. 165 will serve as a reference.

When the pincushion correction tube, V6A, develops a heater-to-cathode short, the 60-Hz (heater voltage) is coupled through the 560-ohm cathode resistor to the vertical output transformer. The effect of this 60 Hz AC would be viewed on the receiver as narrow, dark horizontal lines (6 inches or so, apart) moving slowly upward through the picture.

On some chassis, the cathode of this tube is connected directly to the vertical output transformer. In this case, the 60 Hz AC voltage may be strong enough to modulate the vertical sweep and cause the bottom of the raster to "pull-up" two or three inches and take on a "breathing" action also.

Notice that the unfiltered boost voltage is coupled to the grid of the tube via a 330 pf capacitor (C122). The AC component of this voltage is used to achieve the desired output waveform. A leaky or shorted capacitor (C122) will cause an excessive amount of DC voltage to be applied at the grid and cause extremely high current to be drawn through the tube.

This excessive current will damage the tube, overheat the cathode resistor, if used, and blow the fuse when the tube finally shorts. Check or replace capacitor C122 and the resistor, if used, before replacing the tube and fuse.

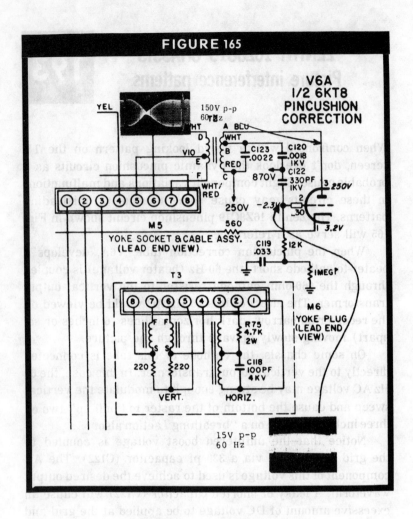

ZENITH 16Z8C19 CHASSIS
Hum bars through the picture

194

This hum appeared on the video signal oscilloscope waveform. The bars seemed to be more noticeable on a blank screen. These are light gray shaded horizontal bars.

The possible cause is a defective (open) zener diode (X16) in the 24-volt power supply low-voltage section. This zener diode not only regulates the 24v supply but also acts as a filter capacitor. (Notice the power supply schematic in Fig. 166.) Use your oscilloscope to check for ripple in the power supply system. The supply voltage must be good if the other chassis systems are to function properly. Also, zener diode X16 will short and burn up R126, a 370-ohm resistor.

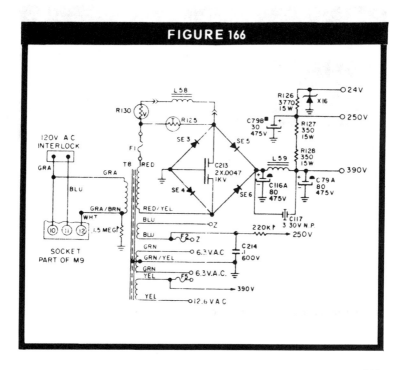

FIGURE 166

195 ZENITH 16Z8C50 CHASSIS
Picture bends and pulls

This color receiver displayed a picture with the "bends" and it appeared as if it were due to an AGC fault. However, the AGC section checked out good.

The oscilloscope was used to check all of the B+ supply voltages, and horizontal pulses were found on the 24-volt bus. The trouble was isolated to a defective (shorted) diode located in the blanker transistor circuit which caused horizontal pulses to appear on the 24-volt bus and also on the AGC transistor collector. Check or replace diode X17, shown in Fig. 167, to correct this trouble.

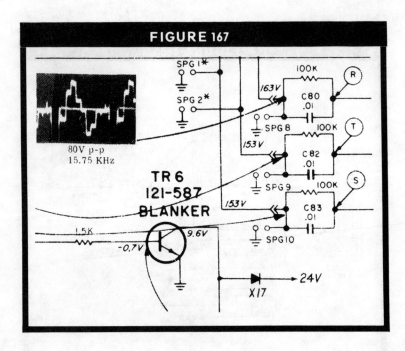

FIGURE 167

ZENITH 16Z8C50 CHASSIS
Drive lines and AGC trouble
196

This same trouble will cause many interrelated symptoms. These symptoms are as follows:

Drive lines at the center of the raster.

AGC control will not operate properly and picture contrast varies from light to dark.

Horizontal and vertical sync will be unstable.

Intermittent color or color hue shift.

May also cause buzz in the audio.

The probable cause for all of these troubles is a defective 6U10 tube. Check or replace the 6U10, since it drives the HV sweep circuitry, keys the burst amplifier to obtain color, controls the 3.58-MHz oscillator, and keys the AGC stage. This tip could save you many hours of needless troubleshooting time. See the schematic of this circuit in Fig. 168.

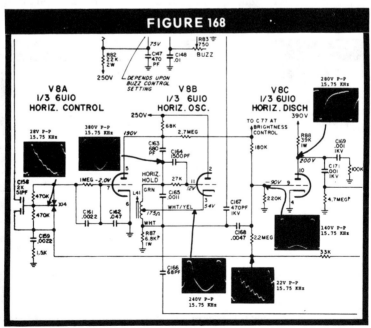

197 ZENITH 20Z1C37 CHASSIS
Vertical shading

This Zenith set had a vertical shading problem on the right side of the screen. This is usually more pronounced as the contrast is turned up. This same trouble may show up as jail bars across the screen.

The probable cause is a defective horizontal-blanking diode. This chassis uses three diodes (one of them is a clamp, see Fig. 169). Some of the other chassis use only two diodes. The horizontal blanking pulse is applied to the video amplifiers, through the diodes to the emitter of the 2nd video transistor amplifier. Thus, turning the contrast down will make the shading "weaker."

Use a scope to check this blanking pulse. The shading will appear as a "slant" of the video signal, with the scope sweep set for the horizontal rate. Also, it's easy to disconnect the diode and check for leakage or replace it. Video amplifier transistor (TR1) leakage may also cause this same shading trouble.

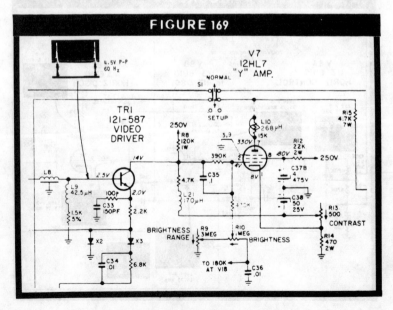

FIGURE 169

ALL BRANDS OF COLOR RECEIVERS
Intermittent circuit connections and circuit board leakage   198

Some color chassis may develop an intermittent condition that will occasionally be caused by an open or cold solder ground connection. In these instances, an oscilloscope can be used as a valuable aid in locating these faulty connections. The oscilloscope is connected to the "high" side of the component connected to ground. Then, while tapping or twisting the chassis or ground post, observe the oscilloscope trace for any change in deflection. The scope's vertical gain should be set as high as practical. In many instances an oscilloscope is more sensitive than an ohmmeter for these poor ground connection applications.

A good way to locate internal PC board leakage is to place a bright light on one side of the circuit board and look at the other side. Should the PC board appear dark or discolored in some areas, those sections should be checked out for suspected circuit leakage.

LOTS OF SNOW ON THE SCREEN   199

This last trouble is not a color TV chassis problem, but was a TV reception condition that I think you will find interesting.

While doing some TV troubleshooting, way up north near Nome, Alaska, a few years ago I received a TV service complaint early one extremely cold winter morning. An old-time prospector came into the shop and as he stomped the snow from his boots, asked me to check his new TV receiver out at his cabin.

Seems as how he couldn't bring in picture or sound that morning on his TV receiver. He also reported: "My set has this here funny looking snow on the screen and it sorta looks like the TV pitcher sender is maybe, outa kilter." "Oh, I said, I'll check with the TV station's chief engineer. Could be the TV transmitter was off the air." A call to the engineer proved fruitless, as he indicated they were transmitting a good picture at full power.

I made this service call late that evening, checking out the TV set and antenna thoroughly. And by all indications they were working properly. But, there was no sound or any indication of a picture on the TV screen. Just lots of snow. However, this snow had a very peculiar appearance because it did not move or dance about. Now if you can just visualize that image in your mind's eye. Just stood still like it was, well, maybe, frozen on the screen. As we stood transfixed, watching the set, suddenly, picture and sound came on clear and strong with the evening news report. But, something was wrong, it was yesterday's news cast. Boy, now wouldn't this tempt you to quit the TV service business.

Well now, it seems the temperature had now warmed up to about 50 degrees below zero, causing the RF signal to thaw out on the TV antenna. It was so cold the TV signal had frozen on the antenna. You might say we had megacicles or kilocicles instead of icicles frozen on the old timer's antenna. Now, wouldn't that just make you chomp your bit.

And another strange thing, this "odd ball" reception problem happened on the first day of April. Very interesting, don't you agree?